Understanding Building Stones and Stone Buildings

Understanding Building Stones and Stone Buildings

John A. Hudson† and
John W. Cosgrove
*Department of Earth Science and Engineering,
Imperial College London, UK*

CRC Press
Taylor & Francis Group
Boca Raton London New York Leiden

CRC Press is an imprint of the
Taylor & Francis Group, an **informa** business
A BALKEMA BOOK

The front cover picture (and Figure 8.7) is of a house built into the ruins of one of the three west front arches at the centre of the original abbey church at Bury St. Edmunds in Suffolk, England, see also the sketch on the back cover.

CRC Press/Balkema is an imprint of the Taylor & Francis Group, an informa business

Typeset by Apex CoVantage, LLC

Library of Congress Cataloging-in-Publication Data
Names: Hudson, J. A. (John A.), 1940–2019 author. | Cosgrove, J. W. (John W.), author.
Title: Understanding building stones and stone buildings / by John A. Hudson and John W. Cosgrove, Department of Earth Science and Engineering Imperial College London, UK.
Description: First edition. | Leiden, The Netherlands : CRC Press/Balkema, [2019] | Includes bibliographical references and index.
Identifiers: LCCN 2018061291 (print) | LCCN 2019000359 (ebook) | ISBN 9781315100180 (ebook) | ISBN 9781138094222 (hardcover : alk. paper)
Subjects: LCSH: Building, Stone. | Stone buildings.
Classification: LCC TH1201 (ebook) | LCC TH1201 .H826 2019 (print) | DDC 624.1/832—dc23
LC record available at https://lccn.loc.gov/2018061291

Published by: CRC Press/Balkema
Schipholweg 107c, 2316 XC Leiden, The Netherlands
e-mail: Pub.NL@taylorandfrancis.com
www.crcpress.com – www.taylorandfrancis.com

ISBN: 978-1-138-09422-2 (hbk)
ISBN: 978-1-315-10018-0 (eBook)
DOI: https://doi.org/10.1201/9781315100180

MIX
Paper from responsible sources
FSC® C013056

Printed and bound in Great Britain by
TJ International Ltd, Padstow, Cornwall

Frontispiece

How they came thither, or from whence, no stones of that kind being now to be found in any part of England near it, it is still the mystery, for they are of such immense bulk that no engines, or carriages which we have in use in this age could stir them.

Daniel Defoe, *A Tour Through England and Wales* (1724–6)

One of the Stonehenge sandstone trilithons on Salisbury Plain in England, illustrating an early use of stone, *ca*. 4,500 years ago. The Stonehenge site was donated to the nation by Cecil and Mary Chubb in 1918.

Dedication

This book is dedicated to Carol Hudson, constant companion on building stone field trips, manager of multiple manuscript components, and expert spotter of text and figure errors.

Carol Hudson in front of a modern gabion[1] flint wall at the Roman Verulamium site in St. Albans, England.

1 A gabion is a wire cage filled with stones.

Contents

Preface

This book is about the building stones and stone buildings of Great Britain, i.e., England, Scotland and Wales. The basis of discussion concerning the building stones must be the geology of the regions because this dictates which types of building stones are locally available, although nowadays the stones can be transported considerable distances as required. Luckily, Britain is endowed with an extraordinary range of different rock types which have been laid down during the billions of years of the geological time spectrum, each rock type having different physical and chemical properties and hence different qualities. In addition to the rock types, we discuss a variety of other factors, e.g., geographical, historical and architectural issues, so that the building stones can be considered 'in context'—from their quarry extraction to their eventual deterioration.

The motivation for writing the book has been the fact that life is much richer the more one understands about one's surrounding environment, whether it be the natural or the built environment. The identification of a particular bird or flower on a country walk provides an extra measure of satisfaction, and so does the identification of a particular building stone together with its architectural context. While writing about the various subjects in the book, i.e., the geology, the mechanics and the architecture, we have tried to ensure that the explanations can be directly understood by everyone, so it is not necessary to have any previous knowledge of the topics when reading the book.

Although the book is primarily about the building stones and stone buildings in Great Britain, occasionally we have also included photographs and text relating to other countries where these help to explain a particular point. All the photographs in the book have been taken by the authors, except for a few which have been otherwise attributed.

John A. Hudson and John W. Cosgrove
Professors, Department of Earth Science and Engineering, Imperial College, London

John Hudson was an academic of international standing in the world of rock mechanics and rock engineering, and a prolific author in this field. He was greatly admired by his colleagues and students for his wisdom, patience and kindness. This book represents the culmination of a lifetime's interest in the architecture of historical stone buildings, the stones they were built from and how their form evolved over time. His passion for the open country, particularly hilly places, merged with an appreciation for the way cathedrals and castles had been inserted to best advantage within the terrain. His heart, soul and love of books, especially old ones, all combined to result in this, his last manuscript, which he completed shortly before his death. He is sadly missed by all who knew him.

—John Cosgrove,
March 2019

Chapter 1

Introduction

1.1 PURPOSE AND CONTENT OF THE BOOK

Our ambition in writing this book is to enhance the reader's understanding and hence appreciation and enjoyment of building stones and stone buildings. Stone has been used structurally and decoratively for many centuries throughout the world—and stone has its own special appeal because of its natural occurrence, its strength and its surface appearance. Today, many old stone buildings still survive, intact or in ruins, depending on their resistance to the many factors contributing to their degradation. Moreover, stone continues to be used extensively as decorative cladding to new and architecturally dramatic buildings.

As an introductory example of a stone building, in Figure 1.1 we illustrate the Jewel Tower in Westminster, London which, as its name implies, was built in 1365 as a secure repository for Edward III's jewels. The Tower, built with Kentish ragstone (a type of limestone), has survived to this day, despite the 1834 fire which destroyed the adjacent Houses of Parliament, and it is now part of the Westminster World Heritage Site. Given that buildings like this, built mainly or partly from quarried stone plus more modern ones with stone cladding, are all around us, the purpose of this book is to present an overview of the whole subject together with the key related supporting subjects in order to enhance the reader's appreciation of building stones and stone buildings. It is a fascinating topic—rich in history, geography, geology, mechanics, chemistry and architecture—and it is our objective to present and explain the related information in a simple, structured, coherent and illustrated way.

There has been a long history of stone use from the Paleolithic era, through the Mesolithic, Neolithic, Bronze Age, Iron Age and Roman times, plus the relatively more recent Anglo-Saxon, Medieval, Post-Medieval and Industrial periods right up to the present day. Stone, especially flint, has been used extensively for tools and buildings from the early Ages. In both our Frontispiece and Figure 1.2, the megalithic nature of Stonehenge is illustrated—one of the most iconic Neolithic monuments in Europe. In Figure 1.3 a detail of the use of flint for constructing the walls of a Roman bathhouse in England is highlighted and in Figure 1.4 a portion of Trajan's column (constructed to commemorate his victory in the Dacian wars) shows Roman wall building. In Figure 1.5 the more recent and extraordinary skills of the 15th century Inca masons are demonstrated, particularly in preparing such perfectly interlocking stone blocks for their dry stone walls in Peru.

Building stones are generally obtained from quarries, although there are exceptions such as stones found on moors and the use of stones from ruined buildings, as was the latter case in Hertfordshire, England, where flints and tiles from the abandoned Roman town of

(a) The Jewel Tower (built *ca.* 1360s) to house Edward III's treasures.

(b) Rough Kentish ragstone to the right of the entrance. This limestone has been used extensively in the SE of England.

Figure 1.1 The Jewel Tower located near the Houses of Parliament, Westminster, London.

Figure 1.2 The 4,500-year-old late Neolithic Stonehenge sandstone monument located on Salisbury Plain, Wiltshire, England (see also the Frontispiece).

Figure 1.3 Portion of a flint wall in a Roman Bathhouse, Welwyn, England (~0.7 m width).

Figure 1.4 Roman brick/stone building, as shown on Trajan's Column completed in AD 113.

Figure 1.5 The interlocking dry stone walls of Sacsayhuaman, an ancient Inca fortified complex constructed of huge, megalithic stones, some weighing over 100 tons. They fit together with an extraordinary precision (a sheet of paper cannot be slid between them) and it is still not known how these limestone blocks were cut. (Courtesy of N. R. Barton)

Verulamium were scavenged for use in the construction of St. Albans Abbey. In the case of quarries for building stone, as opposed to open pit mines excavated for metallic ores, financial considerations generally dictate that the rock stratum being exploited in a quarry should be relatively near the surface and easily accessible. Moreover, the quarry should ideally be located near to where the stone is to be used, although nowadays this latter criterion is not so critical, especially for decorative stones that tend to have higher values and hence can be transported further, as indeed is the case for the popular Carrara marble from Italy which we highlight later in Chapter 4.

So our story should begin with the origin of stone by noting geological aspects that determine both the geography of different types of stone and the associated quarrying process, plus cutting, fashioning and transporting the stone, as illustrated in Figure 1.6 by the granite blocks (known as setts) being used for a market place surface. Once excavated, shaped and used for the exterior of a building or other structure, the stone begins its exposure to the elements and its long and inevitable process of deterioration, decay and journey 'back to the Earth' via the different physical, chemical and biologically associated

Figure 1.6 Granite setts being installed in a road surface.

forms of degradation. Somewhere along the line, an old building stone may be replaced by a newer version but this process is fraught with problems—a subject which we discuss later in Section 7.3.5.

As noted in the Preface, while walking in a countryside environment it is satisfying to be able to name a particular species of bird or flower that one passes, and the same applies to the identification of the different types of building stones in the built environment. Some of these stones are instantly recognisable from their texture, such as the Rapakivi granite from Finland with its characteristic appearance (geologically known as ovoidal, orthoclase phenocrysts), Figure 1.7, and which can sometimes be seen as a decorative stone on the exterior of buildings and often on the counters in Caffè Nero coffee outlets. Another easily recognisable stone is Yorkstone, Figure 1.8. So, following Chapter 2, which provides an overview guide to the geological origin of building stones and associated subjects, Chapter 3 is devoted to the descriptions and illustrations of granites, volcanic stones, limestones, sandstones, flint, metamorphic stones, breccias and conglomerates, plus a variety of artificial stones: Coade stone, terracotta, faience (glazed terracotta), brickwork and concrete. Although most of these latter materials do not have the same character as natural stone, they certainly do have various economic, decorative and durability advantages.

Some relatively less used building stones can have clear visual characteristics, but not be so easily identified when used some distance from their origin; an example of this is the Runcorn sandstone from the west of England used for All Saints Church in Hertford in the east of England, Figure 1.9. However, we hope that the guidance given in this book will

Figure 1.7 The texture of Finnish Rapakivi granite, widely used as a decorative stone (sample ~0.3 m width). The word 'Rapakivi' in Finnish means 'weathered' or 'rotten' due to its appearance, but the granite is actually significantly resistant to weathering.

Figure 1.8 Yorkstone (a sandstone) used for pavements and road construction, Rochester, England.

Figure 1.9 The Runcorn sandstone exterior of All Saints Church, Hertford, England. (Stone bedding texture photographically accentuated.)

enable recognition of the type of such building stones, if not the actual name of a particular variety.

Chapter 4 describes the life of a building stone from the initial quarrying and preparation through to its eventual decay and deterioration 'back to the Earth'. A map is included showing where the different geological types of building stones occur in Britain followed by explanations of the process of quarrying, with notes on how the naturally occurring rock fractures can affect this process. Because of the weight of stone and the lack of suitable transportation methods, in earlier days the stone buildings were built using local stone—so that different types of 'vernacular architecture' developed as a result of a local stone's properties. For example, in the west of England in Cornwall and where there are several large granitic outcrops, a particular local style is evident; whereas on the other side of England in the east where building stone *per se* is either absent or too soft, there is a different local style, e.g., where church towers are often made of flint and are circular—because of the lack of sufficiently strong, larger stones to form the corners, the quoins, of rectangular towers. In this chapter, we also highlight and describe the most famous stone quarry complex in Britain—the Portland limestone quarry system in Dorset—and the most famous stone quarry complex in Europe—the Carrara marble quarries in Italy.

This leads in Chapter 5 to a discussion of stone buildings, from pillars, lighthouses and walls, through arches, bridges, buttresses and roof vaults, to castles and cathedrals. In particular, we highlight the extraordinary achievement of Alan Stevenson in building the Skerryvore lighthouse off the west coast of Scotland in Victorian times and we note the different appearances of dry stone walls as a function of the stone used. It is easy to understand the stability of stone walls but it is not so intuitively obvious why stone arches can be so stable that they can remain in existence for thousands of years. We explain the mechanics of such arches, provide examples via the three stone bridges across the tidal region of the river Thames, and extend the discussion to wall buttresses and the stability of the spectacular stone roof vaults in many cathedrals. The overall architectural nature of castles and cathedrals is also discussed.

The remainder of Chapter 5 is devoted to stone lettering. At various times in the past, mankind has used lettering and carving in stone to record information of different types; indeed, it is mainly through these messages that we are able to receive contemporaneous information from the past, often thousands of years old. Such messages have been inscribed in individual stones and in stone buildings, a practice that continues today. Thus, the subject of lettering in stone is a particularly important one in the context of this book, and so we discuss cuneiform and other types of historical alphabets, in particular the Roman alphabet which persists to this day for prestigious inscriptions. Nowadays, stone lettering can be exceptionally precise through the use of computer guided laser and abrasive cutting techniques.

Following Chapter 5 describing the stone buildings themselves, Chapter 6 discusses the evolution of architectural styles, from Saxon times through to post-modern buildings. We explain that this evolution occurred directly as a result of mechanical principles: firstly, using stone lintels across two uprights, i.e., trilithons like the Stonehenge example in our Frontispiece, then the use of the semi-circular arch, then the use of the pointed arch, plus the subsequent use of more modern techniques. We expand this simple classification into a description of the main sequence of styles: Egyptian, Greek and Roman, and then the later styles as related to the early religious stone buildings in Britain: i.e., Saxon, Norman, Early English, Decorated and Perpendicular. We also include an explanation of post-modern architecture. The history of architecture is indeed the history of civilisation written in stone. Chapter 6 concludes with summaries of the extraordinary contributions made to architecture by the Roman Marcus Vitruvius Pollio and the British Sir Christopher Wren, plus a list of books containing explanations and glossaries of architectural terms.

In order to illustrate many of the points made in the book, more comprehensive descriptions of two exemplary stone buildings are provided in Chapter 7: these are of the Albert Memorial in London (Fig. 1.10) and Durham Cathedral (Fig. 1.11) in the north of England. The first is chosen to illustrate a superb *decorative stone* structure, together with the wide range of stones that were used; and the second is to illustrate a masterly *building stone* structure.

The Albert Memorial is located in Kensington Gardens in London and commemorates the life of Prince Albert, the husband of Queen Victoria. His death from typhoid fever in 1861 had a profound and prolonged effect on Queen Victoria. As Prince Consort he had been active in many areas and, in particular, he was a prime organiser of the Great Exhibition of 1851. So, it was decided that a commemorative structure should be prepared and the architect George Gilbert Scott won the competition to design an appropriate tribute. The resulting spectacular Memorial, unveiled in 1872, is described as a 'high-Victorian Gothic extravaganza' and indeed contains a wealth of different decorative building stones sourced from around Britain—which is why we have chosen it as our first exemplary stone structure. Many interesting details relating to the geological aspects of the different decorative stones used and the architectural style of the Memorial are described in the first part of Chapter 7.

Durham Cathedral, Figure 1.11, in the north of England is famous for its Norman architecture, for the introduction of the 'Gothic arch', and for the stone roof vaulting. The stone used for the Cathedral's construction is a local sandstone which has weathered significantly

Figure 1.10 The Albert Memorial, London, built and decorated with a variety of building and decorative stones sourced from different locations.

Figure 1.11 The majesty of Durham Cathedral, built of local sandstone between 1093–1135.

over the last 900 years; so in this second part of Chapter 7 and as a case study in stone weathering, we take the opportunity to describe and illustrate in detail the chemical and mechanical features of the sandstone weathering process, plus the origin of the mysterious 'liesegang rings'. These rings are also visible in Yorkstone, which is another sandstone used ubiquitously as pavement stone throughout Britain.

Following the initial discussion of building stone degradation in Chapter 4 and the weathering of the Durham Cathedral sandstone in Section 7.3, in Chapter 8 we explore more fully the subject of the breakdown of a building stone's microstructure over time and the physical and chemical processes involved. Sometimes the erosion of the stone is relatively simply understood, as in the Bank of England's Portland stone case illustrated in Figure 1.12 where the stone has been exposed to the British weather for many years, especially the rain (dilute carbonic acid) and the frost. Important stone changes can also be induced by atmospheric gases, e.g., sulphur dioxide and nitrogen oxides, causing the formation of soluble gypsum (hydrated calcium sulphate). Also, in Chapter 8 and on the larger scale, we discuss the causes and inevitability of stone building ruins and how, through modern technology, it will be possible for old buildings to be 'reborn'.

One of the difficulties associated with building stone weathering is the replacement of badly deteriorated building stones with new stones, either from the same geological formation, from a similar formation or indeed from a completely different formation. Often, new building stones are visibly different to those of the original building, thus having an adverse aesthetic effect. There is no clear solution to this problem because even substitute building stones from exactly the same geological strata look new and out of place. This and other related deterioration aspects are also discussed in Chapter 8.

Figure 1.12 Weathering of the Portland stone exterior wall of the bank of England in London which has accentuated the plentiful fragmentary fossils.

Another difficulty can be understanding the reason for a particular form of deterioration. As another case study, we discuss Carrara marble from the quarry in Italy where Michelangelo sourced his marble for sculptures. This example is particularly interesting, from both the stone alteration aspect and the stone substitution aspect. The white marble, in the form of cladding slabs, has been used to cover complete buildings—in the example discussed it is Finlandia Hall in downtown Helsinki. After the marble slabs were attached to the building, they began to bow, i.e., become curved. The exact reason for this phenomenon, which is not unique to Finlandia Hall, is uncertain, although it is thought to be related to the high residual stress remaining in the marble from its original tectonic formation, plus the high winter-to-summer temperature differences in Helsinki. In an attempt to solve this problem, a major research effort was undertaken from 2000–2005 by a consortium of 17 organisations under the auspices of the European Community. We provide a summary of this research programme together with a discussion of its conclusions in Chapter 8.

The chapter finishes with a discussion on the inevitability of stone building ruins, see the evocative example in Figure 1.13. Should an attempt be made to preserve such ruins, possibly at significant expense, or should nature take its course and the stone be allowed to decay and revert naturally 'back to the Earth'?

Chapter 9, the concluding chapter, contains a reminder of the key points in the book, as related to our key objective of making the subject of building stones and stone buildings

Figure 1.13 A ruined section of the original sandstone Church of St. John, Chester, England.

Table 1.1 Structure and content of the book

Chpt.	Title	Subjects included
1	Introduction	Explanation of the purpose of the book
2	The geological origin of building stones	Introduction to the geological background
3	Recognising the different types of building stone	Granites, volcanic stones, limestones, sandstones, flint, metamorphic stones, breccias and conglomerates, plus artificial stones
4	The life of a building stone	The complete life history from the quarry source through to decay and deterioration
5	Stone buildings	Pillars, lighthouses, walls, arches, bridges, buttresses, roof vaults, castles, cathedrals; discussion of stone lettering
6	The architecture of stone buildings	An explanation of the main historical architectural styles and their geometrical features, the contributions made by Vitruvius and Christopher Wren, plus notes on modern architecture
7	Two exemplary stone structures	Descriptions of two exemplary stone structures: 1. The Albert Memorial (building stones); 2. Durham Cathedral (stone building)
8	Deterioration of building stones and stone buildings	Mechanisms of deterioration; the inevitability of ruins, digital recording of building geometry and the possibility of building 'rebirth'; case history of the bowing of Carrara marble cladding
9	Concluding comments	Summary, with highlights of the book contents
References and bibliography: a list of the referenced books, plus other books of interest		
Index		

more accessible to those readers who have felt some curiosity when seeing the way in which granites, sandstones, limestones and other rock types have been used in both the structure and decoration of our stone building heritage.

A reminder of the chapter headings and their content is presented in Table 1.1.

In terms of access to further information, we list key internet sites and include a combined references and bibliography section. The references relate to books specifically referred to in the text, whilst the bibliography contains other books of interest. There is a wealth of information, both general and technical, that can be sourced via the internet, plus many books are available through new and used online book outlets, some with evocative titles such as

- *England's Chronicle in Stone* written by James F. Hunnewell in 1886,
- *The Stone Skeleton* written by Jacques Heyman in 1966, and
- *. . . isms: Understanding Architecture* by J. Melvin in 2005.

Chapter 2

The geological origin of building stones

The Earth was made so various, that the mind
Of desultory man, studious of change
And pleased with novelty, might be indulged.
William Cowper (1731–1800)

2.1 INTRODUCTION

The subject of medicine is supported by a knowledge of biology and the subject of engineering by a knowledge of mechanics. Similarly and most importantly, the understanding and enjoyment of studying building stones is greatly enhanced by a basic knowledge of geology. It is not necessary to understand the fundamentals of geology in order to appreciate building stones but, because geology is the foundation stone of the subject, a basic knowledge does provide a much better understanding. It enables the observer to be able to enjoy not only a particular stone building but also the surrounding scenery within its geological context.

Accordingly, in this chapter we provide a brief introduction to geology as it supports the understanding of building stone types and their characteristics. There are many subjects within the whole geological spectrum, so the presentations in this chapter just concentrate on those key features which enable the origins and properties of the building stones to be understood and recognised. Fearnsides and Bulman (1950) made the comment that every kind of fully or partly cemented or consolidated rock, igneous, sedimentary and metamorphic, has somewhere been used as a building stone. Nevertheless, most building stones can be simply described as granite, limestone and sandstone, but these broad categories will contain different types having varying physical and chemical properties as noted in the following chapter sections. In other words, this chapter introduces the reader who may be unfamiliar with Earth Sciences to the various aspects of geology that will help in understanding the mechanical and decorative properties of building stones.

Included are introductions to the classification of minerals and rocks, the aspect of time in geology, the structure of the Earth and plate tectonics, leading to the distribution of different types of rocks in the UK and hence the *in situ* locations of the building stones and consequently different types of local architecture. The chapter ends with supporting photographs of stone buildings from various geological locations in Britain.

We note that the *Oxford English Dictionary* defines stone as "Hard, solid, non-metallic, mineral matter of which rock is made, especially as a building material." Etymologically, the word 'stone' derives from the Old English 'stān' (noun) which is of Germanic origin and related to the Dutch 'steen' and German 'stein'. Stones are individual pieces of rock, generally the result of the inexorable breakdown of rocks exposed at the Earth's surface by the processes of weathering and erosion, i.e., they are pieces of rock detached from their parent rock mass. These 'pieces' occur on a large range of scales in nature from minute grains to blocks kilometres in size. Some of these naturally occurring stones can be used 'as is' for building stones. However and more commonly, building stones are quarried, i.e., detached from the *in situ* rock mass, often by exploiting natural fractures, and then 'dressed' (the shaping and surfacing of blocks of stone) as required for the construction of buildings. The relation between the rock mass type and the mode of quarrying is discussed in Chapter 4 after introduction to the specific types of building stone in Chapter 3.

In the current geological context defining a stone as a piece of rock begs the question, "what is a rock"? Although all readers will be familiar with rock, its definition is: "a solid aggregate of one or more minerals or mineraloids, the latter being mineral-like substances that do not demonstrate crystallinity and which can possess chemical compositions varying beyond the generally accepted ranges for specific minerals." (For example, obsidian is an amorphous glass and not a crystal, generated when a granitic magma erupts onto the Earth's surface and chills too rapidly for crystals to form.) Thus, in order to fully understand building stone, it is first necessary to be able to relate to the classifications of minerals and rocks. For this reason, these are outlined and discussed in the following sections.

2.2 CLASSIFICATION OF MINERALS

An early definition of a mineral was given by Friedrich Mohs (1773–1839), a German scientist, who defined minerals as 'inorganic products of nature'. We can amplify this definition to 'a mineral is a naturally occurring, solid, homogeneous, crystalline, chemical element or compound'. Mohs is most famous for his scale of hardness based on the ability of one mineral to scratch another; it is a measure of the resistance of the mineral to scratching, noting that the hardness of a mineral is mainly controlled by the strength of the bonding between the atoms and the size of the atoms. There are ten minerals in Mohs scale: talc (the softest), then gypsum, calcite, fluorite, apatite, orthoclase, quartz, topaz, corundum and diamond (the hardest), Figure 2.1. The logic of the Mohs scale is that any mineral can scratch any mineral below it on the scale, and conversely can be scratched by any mineral above it.

Because Mohs' classification of minerals is based on one physical property, i.e., hardness, it can be of use when considering the mechanical strength of a rock (an agglomeration of minerals) in terms of its suitability as a building stone. However this classification was formulated prior to the enlightenment provided by research in the late 19th and 20th centuries when the chemical composition and the structure of crystal lattices became better understood. Mohs' definition of a mineral as 'an inorganic product of nature' was therefore modified to a mineral is 'a naturally occurring inorganic solid with a definite chemical composition and an ordered atomic arrangement'.

Although Mohs scale of hardness remains useful, especially in field studies, it cannot be considered to be a comprehensive scheme for mineral classification as it involves only 10 minerals of the more than 3000 which occur in nature. The classification currently used is that proposed by Professor James Dana of Yale University in 1848. So, we can now define a rock as a naturally occurring solid made up of an aggregate of one or more minerals. For example,

- limestone is composed of only one mineral: calcite,
- basalt is commonly composed of three main minerals: feldspar, pyroxene and olivine,
- granite is commonly composed of five minerals: two kinds of feldspar, mica, amphibole and quartz.

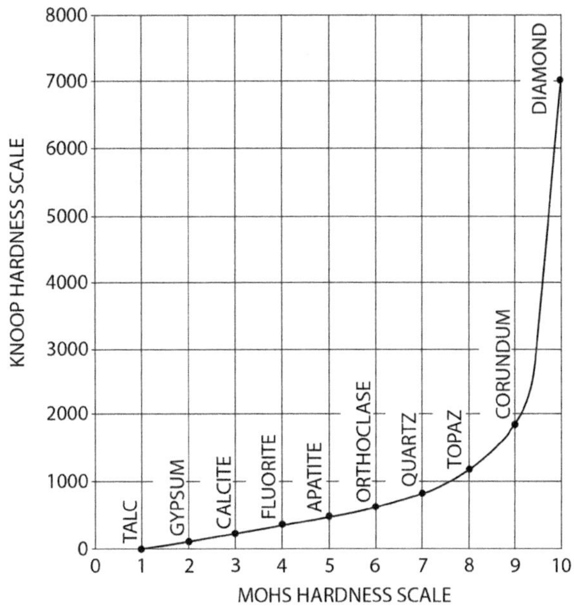

Figure 2.1 A graphical representation of Mohs scale of mineral hardness with the type minerals indicated for each value and the relation with the alternative Knoop hardness scale.

This leads us on to the classification of rocks and hence building stones as described in the next section.

2.3 CLASSIFICATION OF ROCKS

Geologists divide rocks into three groups: igneous, sedimentary and metamorphic, Figure 2.2.

- **Igneous** rocks crystallise from magma (hot molten rock which forms in the mantle and the lower part of the Earth's crust, Fig. 2.5).
- **Sedimentary** rocks form by weathering and erosion of pre-existing rock to make sediment, which is then lithified (cemented) into rock.
- **Metamorphic** rocks form by the deformation and/or recrystallisation of pre-existing rock by changes in temperature, pressure and/or chemistry.

All rocks can be placed into one of these three groups.

With reference to Figure 2.2, some of the other important terms used are as follows. *Clastic* applies to a rock composed principally of broken fragments (i.e., clasts). *Intrusive* applies to rocks formed by the injection and cooling of magma **within** the Earth; *extrusive* applies to rocks formed by the extrusion and cooling of magma (lava) **onto** the Earth's surface; *foliated* means that the structural and textural features of a rock have a similar orientation, either as a result of sedimentation or deformation. For example, a slate can be split easily into thin sheets as a result of the micas within it having been aligned parallel to each other during tectonic compression.

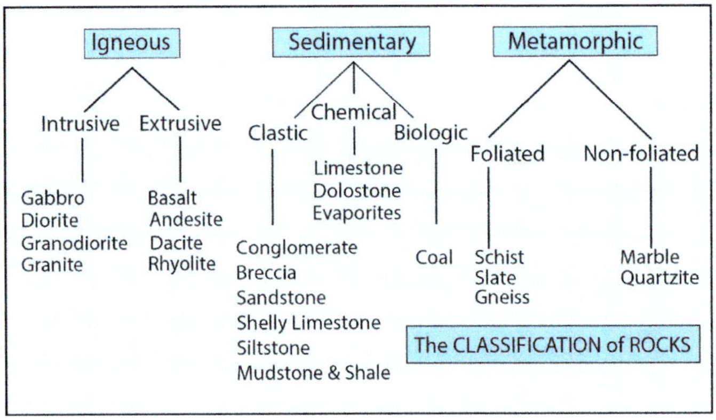

Figure 2.2 The classification of rocks. (Adapted from an original courtesy of Steve R. Mattox)

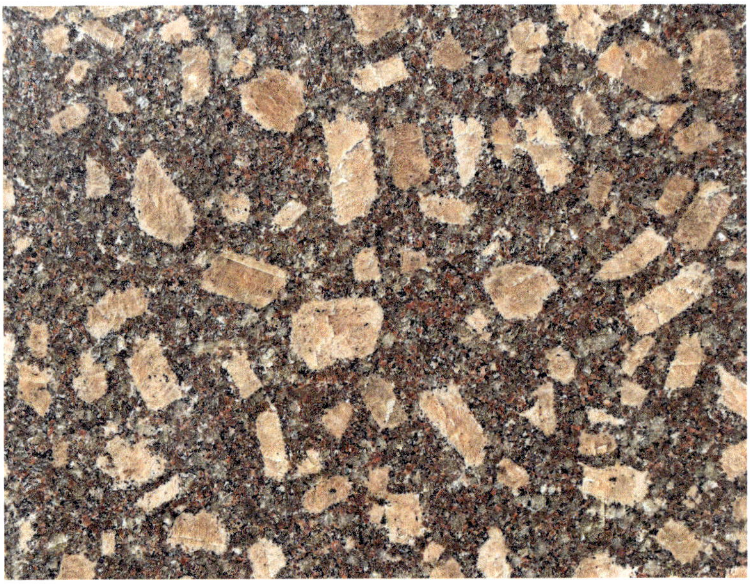

Figure 2.3 Large crystals (phenocrysts) of feldspars in an intrusive rock, a granite. Shop façade, Piccadilly, London.

Igneous rocks (primary rocks) form by the cooling of magma and are divided into those that *intrude* into the Earth's crust where they cool and solidify and those which *extrude* onto the earth's surface as lava flows. The main differences between the intrusive (plutonic) and extrusive (volcanic) rocks is that the intrusive ones cool slowly—with the result that there is time for large crystals to grow, Figure 2.3; whereas, the extrusive ones cool rapidly and, as a consequence, are much finer grained.

The rocks are classified on the basis of the composition of the magma from which they cool and the rate of cooling which determines the grain size (magmas being classified via the amount of quartz they contain, Fig. 2.4). The slow cooling of a magma rich in quartz produces a granite (a coarse grained rock); and the rapid cooling, which occurs when the magma is extruded onto the Earth's surface as a lava flow, produces a rhyolite (a fine-grained rock). Sometimes the magma is cooled (quenched) so rapidly that even very small individual minerals have no time to grow: this is how volcanic glass (obsidian) is formed.

Figure 2.4 shows the seven major minerals found in igneous rocks. The different types of igneous rocks plot along the horizontal axis with granite (and its fine-grained equivalent rhyolite) on the left and gabbro (and its fine-grained equivalent basalt) on the right. The composition of the different rocks can be read off by dropping a line vertically through the chart, see as an example the dashed line on the left side of Figure 2.4. This granite is made up of approximately equal amounts of quartz and orthoclase feldspar (about 30% of each), with lesser amounts of plagioclase feldspar (about 15%) plus mica and amphibole. Notice also that the term granite is applied to rocks with a range of composition, as indicated by the arrows at the top of Figure 2.4.

Sedimentary rocks are classified based on their mechanisms of formation: i.e., clastic (broken) rocks, chemical rocks and biological rocks, Figure 2.2.

Clastic rocks are the main type of sedimentary rock and are formed from fragments of other rocks (igneous, sedimentary or metamorphic) and are therefore regarded as secondary rocks. Fragments (clasts) of rock, loosened from their outcrop at the Earth's surface by the processes of weathering and erosion, are transported to some basin or depression by water or wind (e.g., desert sands and volcanic tuffs) where the sediments are trapped. These become

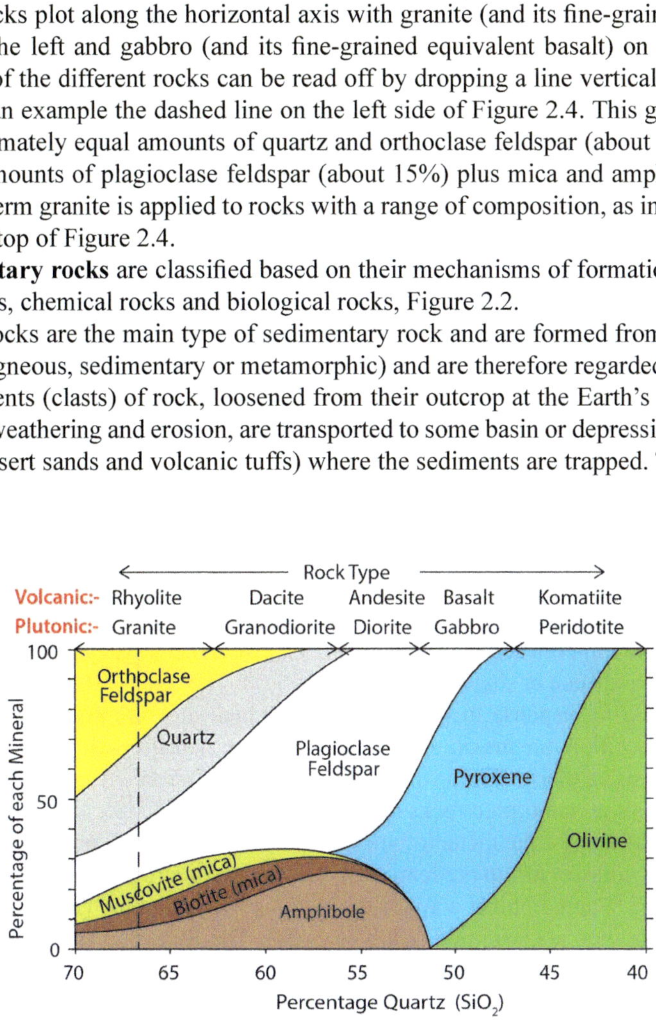

Figure 2.4 The composition of common igneous rocks. Note the reduction of quartz and orthoclase feldspar as one moves from left to right in the diagram. The vertical dashed line indicates how the diagram is used: for any position along the Rock Type axis, the dashed line directly indicates the percentage mineral composition of the rock type.

compacted and cemented as they are buried beneath younger sediments to form sedimentary rocks. As with igneous rocks, they are classified on the basis of

- *Composition*, e.g., sandstone which is composed mainly of quartz grains, and limestone which consists mainly of calcite, often in the form of shells and shell fragments, and
- *Grain size*, e.g., sandstone is classified according to grain size: very fine grains (siltstones); coarser grains (sandstones); and very coarse grained consisting of quartz rich pebbles (conglomerates).

Chemical sedimentary rocks include chert/flint (silica) and some limestones (calcite) which form as a result of precipitation out of sea water or from water within the sediment or rock saturated or over-saturated with one of these minerals. Other common types of chemical rock include the evaporites, e.g., salt (halite which is sodium chloride, NaCl) and gypsum (which is hydrated calcium sulphate, $CaSO_{4.}2H_2O$)—which are precipitated when seas and lakes evaporate.

Biological sedimentary rocks form from once-living organisms. They may, like coal, form from accumulated carbon-rich plant material or, like some limestones, from deposits of animal shells.

Metamorphic rocks are, as their name implies, rocks formed by the alteration of other rocks. These changes occur by the processes of mechanical stressing and/or heating. Like sedimentary rocks, they are secondary rocks and are sub-divided into ***foliated*** and ***non-foliated***. The degree of foliation (orientational texture) that forms is determined by the stress field under which the metamorphism occurs and the mineralogy of the rock.

Most metamorphic conditions involve both temperature and stress, and the minerals of the rock recrystallise in response to these effects. If the stress is non-hydrostatic, i.e., if the three principal stresses that define the stress field are unequal, then any platy minerals in the rock, such as clays and micas, will become aligned—with the result that the rock develops a planar fabric. One of the most well-known examples of such a rock is a slate where the aligned micas allow the rock to be split into fine sheets which are ideal for roofing because of their shape and impermeability, see Section 3.7.1. Note that a slate is formed by the metamorphism of a mudstone or shale which contains clays. These recrystallise into micas (platy minerals) aligned in response to the stress under which they are crystallising, with their planes at right angles to the maximum principal compressive stress. It is this alignment that imparts the pervasive planar fabric which characterises the rock.

The names of metamorphic rocks are based on their grain size, which in turn is determined by the intensity of the metamorphism. Relatively low grade metamorphism changes a shale and mudstone (made up predominantly of clay and quartz) into a slate. Progressively higher grades of metamorphism convert the slate first to a phyllite (intermediate between slate and schist), then a schist, and finally a gneiss. The transition from slate to gneiss is accompanied by an increase in grain size.

In contrast, sandstones, which are made up predominantly of quartz grains, do not form a foliated rock when recrystallised under the same stress field. This is because quartz is not a platy mineral and therefore, even if the individual grains are crystallographically aligned during metamorphism, the resulting rock, a quartzite, will not possess a planar fabric. Similarly, limestone, which is made up of calcite, i.e., not a platy mineral, on metamorphism is converted to a marble which is not foliated. As mentioned earlier, the specific case of Carrara marble from Italy is particularly interesting and is discussed in Sections 3.7 and 4.2.

2.4 THE STRUCTURE OF THE EARTH AND PLATE TECTONICS

Although the structure of the Earth may seem to be too large a scale to be relevant to building stones in Britain, in fact the 'bigger picture' is necessary for understanding the overall structural development and current tectonic circumstances of Britain, i.e., not only the distribution of the geological strata but also the presence of major fractures in the rocks, known as faults, which can affect quarrying and are the reason for earthquakes in the UK—and hence the susceptibility of stone buildings to damage.

The structure of the Earth is shown in Figure 2.5. It consists of a central core made up of iron and nickel, overlain by the mantle which is made up predominantly of iron magnesium silicates (pyroxene and olivine, see Fig. 2.4), and the crust which is divided into the oceanic and continental components, the former being composed mainly of basalt and the latter being composed mainly of quartz (silica), feldspars and some iron magnesium silicates.

A weak layer in the mantle, the asthenosphere shown in Figure 2.5, allows the mantle and crust above it (known collectively as the lithosphere) to become detached from the underlying mantle. Plate tectonics, the process that dominates the deformation of the Earth's crust, is the result of the horizontal movement of various sections of the lithosphere, known as the tectonic plates (see Fig. 2.6), with respect to each other. When the plates move apart, they form oceans characterised by a major, central fracture zone within which large quantities of basaltic magma rise to produce mid-oceanic ridges. These are the divergent plate boundaries shown on Figure 2.6. When plates move towards one another (the subduction zones shown in Fig. 2.6), the resulting collision produces mountain belts and large quantities

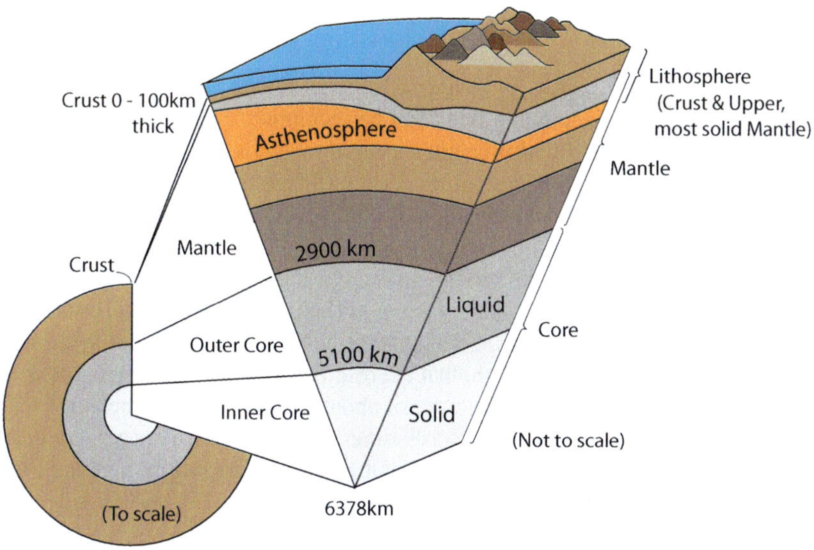

Figure 2.5 The layered structure of the Earth showing the metallic inner and outer core (iron and nickel), the mantle consisting of heavy, iron magnesium silicates, and the crust which is divided into oceanic crust (relatively dense, made up predominantly of basalt) and continental crust (lower density, made up predominantly of quartz, feldspar and iron magnesium silicates). (Modified from "Inside the Earth", U.S. Geological Survey)

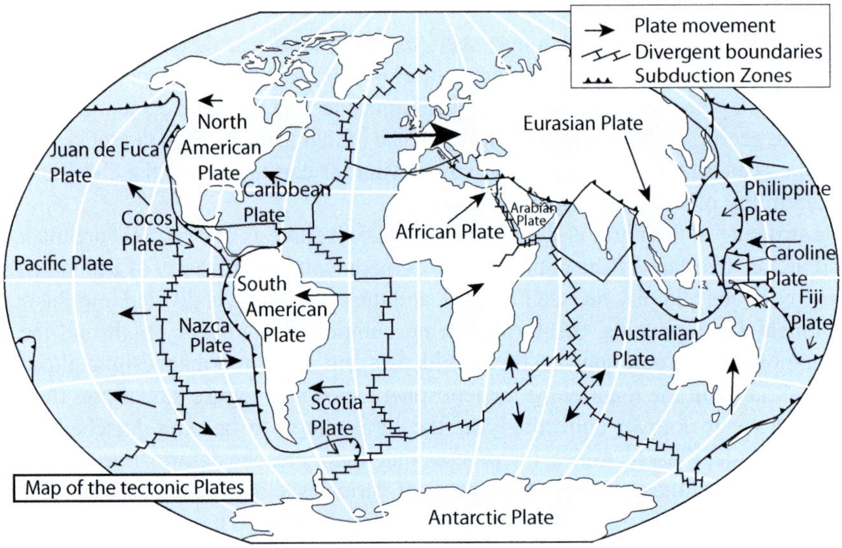

Figure 2.6 The distribution of tectonic plates on the Earth's surface. As the plates move with respect to each other, major fracturing and faulting occurs which results in intense seismicity and the generation of earthquakes along the plate margins. The arrows indicate the relative directions of plate motion; note especially the arrow beneath the British Isles indicating the West–East movement generated by opening along the Mid-Atlantic Ridge. (Adapted from Eric Gaba for Wikimedia Commons (CC by SA 3.0 https://creativecommons.org/licenses/by-sa/3.0))

of granitic magma. Ancient mountain belts can be recognised by these associated belts of granites, and, as discussed later in this chapter, two such belts occur within Britain and these have provided building stones for some of our most prestigious buildings. The plate margins (Fig. 2.6) are regions of the Earth where major faulting occurs and are therefore major zones of seismicity and earthquakes.

In Figure 2.6, note the larger horizontal arrow starting just below the British Isles; this indicates that the Eurasian tectonic plate is being pushed eastwards, or more strictly south-eastwards. The submarine Mid-Atlantic Ridge is formed and the Atlantic is widened, albeit by a small amount each year. It is thought that the resulting stress in the British rocks is about 25 MPa (megapascals)—which is of the order of about 3600 psi (pounds per square inch) or about 1.8 tons per square inch, i.e., roughly 100 times the pressure in a car type, or the weight of a car on each square inch within the rock—but acting horizontally. There are also other stress components in the other perpendicular directions.

2.5 ROCK FRACTURES

Rocks exposed at the Earth's surface are almost invariably fractured, which may be either a help or a hindrance when attempting to quarry building stones. Fractures form in the crust as the result of applied stresses over geological time: they may form early in the rock's history during its burial, later during various tectonic events that affect the rock over geological time, or during the exhumation of the rock which brings it back to the Earth's surface. These fractures fall into two main types: namely, **joints** and **faults**, Figure 2.7. These differ fundamentally in the movement which occurs along them.

Very little movement occurs during the formation of joints and this movement is at **right angles** (90°) to the fracture plane. Such fractures often form in straight, regular arrays known as 'joint sets' and rocks are frequently cut by more than one set of joints. This results in the rock being divided into blocks, Figure 2.7(a), which can aid considerably in the quarrying

(a)

(b)

Figure 2.7 (a) Joints exposed on a Jurassic limestone bedding plane at Lilstock on the Bristol Channel coast, England. The rock surface is 2 m wide. Four joint sets can be seen: one running E–W, i.e., across the page, a less continuous set running N–S, a through-going set running WNW–ESE and another set trending ENE–WSW. These joints combine with the bedding planes to divide the rock into blocks. (b) Two small inclined faults cutting sandstones and shales of Carboniferous age at Northcote Mouth, north of Bude, Devon in the UK. It can be seen that even the relatively small movement along the left-hand fault has caused damage to the adjacent rock. (Photo by the authors, Courtesy of Imperial College Press, London)

process. An impressive example where joints have been used in the extraction of a stone block is shown in Figure 2.8, a view of an ancient 'granite' quarry at Aswan in Upper Egypt. A set of vertical joints running parallel to the obelisk 'which has been carved from the block of granite between two adjacent joints' has been exploited. The unfinished obelisk has been left *in situ* because, regrettably, it cracked during its extraction—probably as the result of an earthquake.

Movement associated with the formation of faults is dominantly **parallel** to the fracture. This can be seen in Figure 2.7(b), which shows the two faults defining a keystone-shaped block that has moved vertically downwards. Many faults experience multiple episodes of such movement along them—which can damage the surrounding rock mass forming a fault damage zone where the rock is significantly fractured and unsuitable for extraction for building stone. In contrast to joints, faults can hamper the extraction of regular blocks of rock and result in sections of a quarry being uneconomic.

Faults occur on all scales in the Earth's crust, ranging from the small-scale, such as the examples shown in Figure 2.7(b), which are only of the order of a few metres in length and

Figure 2.8(a) View of a 42 m long 'unfinished obelisk' lying on the floor of an ancient 'granite' quarry in Aswan, Egypt. The orientation of the obelisk is controlled by and its excavation facilitated by the main set of extensional joints which are vertical. Several of these are clearly visible on the left and right sides of the photograph. Note the fracture running through the upper part of the monolith which caused the project to be abandoned. For comparison Cleopatra's Needle in London is 21 m, 69 ft high, see Figure 2.8(b). (Photo by Olaf Tausch (2009) "Unfinished obelisk at Aswan, Egypt" from Wikimedia Commons (CC BY 3.0 (https://creativecommons.org/licenses/by/3.0))

Figure 2.8(b) Cleopatra's Needle, Thames Embankment, London, one of three ancient Egyptian obelisks created during the reign of the 18th Dynasty Pharoah Thutmose III, approximately 1450 BC.

where displacements are measured in centimetres, to faults which define plate boundaries and which are several hundreds of kilometres in length. Movement on large-scale faults generates earthquakes and these represent major hazards to stone buildings. As noted in the brief discussion on plate tectonics above, major fault activity in the Earth is focused along the tectonic plate margins shown in Figure 2.6 and, as a result, plate margins define the major earthquake zones.

It can be argued that, because Britain is situated away from current plate margins, it is less susceptible to earthquake damage than other regions in the world. This is true, but Britain is not immune from significant seismicity. Figure 2.9 shows the location of earthquake epicentres in the Straits of Dover situated off the south coast of England. As expected, the earthquakes that have occurred here have been located along faults, but these are not related to the current tectonic plate margin which lies some 1000 km south and coincides with the Alpine mountain belt caused by the collision between the African and European plates. In fact, they are related to an ancient plate collision, the Hercynian collision, which occurred at the end of the Carboniferous period, ~300 million years ago, when land masses Gondwanaland in the south collided with Laurasia in the north to produce a single tectonic plate named Pangea. The ancient mountain belt generated by this collision is called the Hercynian

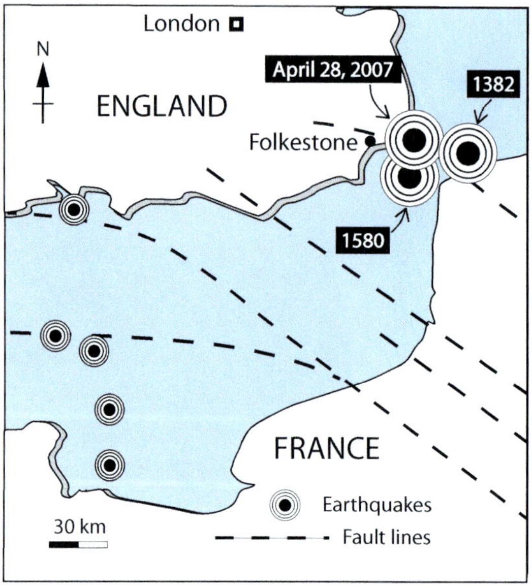

Figure 2.9 The epicentre of the Dover Straits earthquakes is 10 km below the surface of the sea bed. It lies on an ancient fault formed as the result of tectonic plate collision at the end of the Carboniferous period. (Modified from David Derbyshire for *MailOnline* 2010)

Belt; it was as impressive as the present day Alpine-Himalayan Belt and also trended east–west. The faults formed as a result of the Hercynian collision trend E–W and NW–SE, and now represent weaknesses in the crust (Fig. 2.9) which can be reactivated by present-day plate collision—even though they are now situated far from current plate margins.

The 'Great English' Essex earthquake of 1884 centred near Colchester had a Richter magnitude of 4.6 and some readers may have experienced in more recent years small earthquakes in Scotland or in the NW or SE of England. One shook Cumbria in April 2009 with a magnitude of 3.7 on the Richter Scale and was felt widely, and another occurred north-east of Swansea in Wales in February 2018 with a magnitude of 4.4. Geologists expect a magnitude 4.5 quake, i.e., capable of shaking ornaments, at least once a decade in Britain and a moderate 5.5 earthquake, i.e., capable of causing major damage to badly constructed buildings, on average every century. (During the writing of this book, on 9 June 2018 a 3.9 magnitude earthquake occurred below Grimsby in the NE of England.)

2.6 GEOLOGICAL TIME

There are two principal methods for determining the age of rocks. One is radiometric dating (or radioactive dating) which is based on the known decay rate of radioactive isotopes and compares the abundance of a naturally occurring radioactive isotope within the rock to the abundance of its decay products—giving, within error limits, an absolute age. The other is the law of superposition of strata which is based on the assumption that, in general, underlying strata were deposited before the overlying strata, and hence are older. Of course, this latter method of dating provides only the relative ages of the rocks.

The oldest dated Earth minerals (4.0–4.2 billion years old) are small zircon crystals, originally formed in magmas, which are found in the sedimentary rocks of Western Australia. The oldest Moon rocks are from the lunar highlands and were formed when the early lunar crust was partially or entirely molten, dated at between 4.4–4.5 billion years in age. The majority of meteorites have ages of 4.4–4.6 billion years; these meteorites are fragments of asteroids and represent some of the most primitive material in the solar system. The evidence from radiometric dating indicates that Earth is about 4.54 billion years old. Thus, Figure 2.10(a) provides an overall history of the Earth via the clock-type diagram and indicates that the building stones discussed in this book, although millions of years old, were all actually formed relatively late in the Earth's life, i.e., just from 10 o'clock to 12 o'clock. This latter duration has been sub-divided into various eras, periods and epochs of time, as shown in the time charts in both Figures 2.10(a) and (b).

Although super-eons are used when talking about time intervals relating to the age of the universe (currently thought to be ~14 billion years old), the largest unit of time usually used relating to the Earth is the **eon**. Eons are divided into **eras**, which are in turn divided

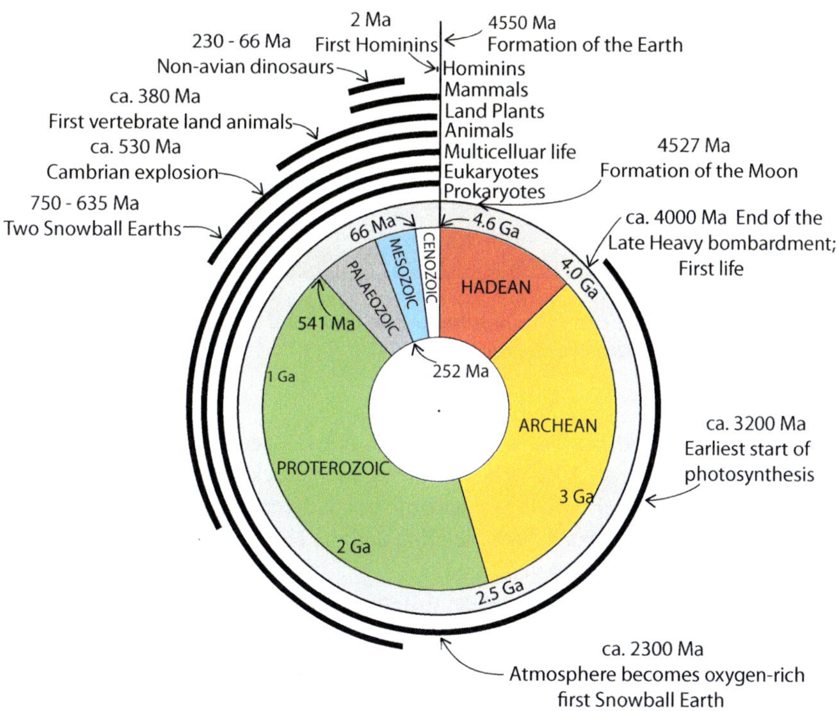

Figure 2.10(a) Geological time represented clockwise with the geological time intervals and important events shown. Note that: Ma refers to millions of years and Ga refers to billions of years; Snowball Earth refers to the time(s) when the Earth's surface was entirely frozen.

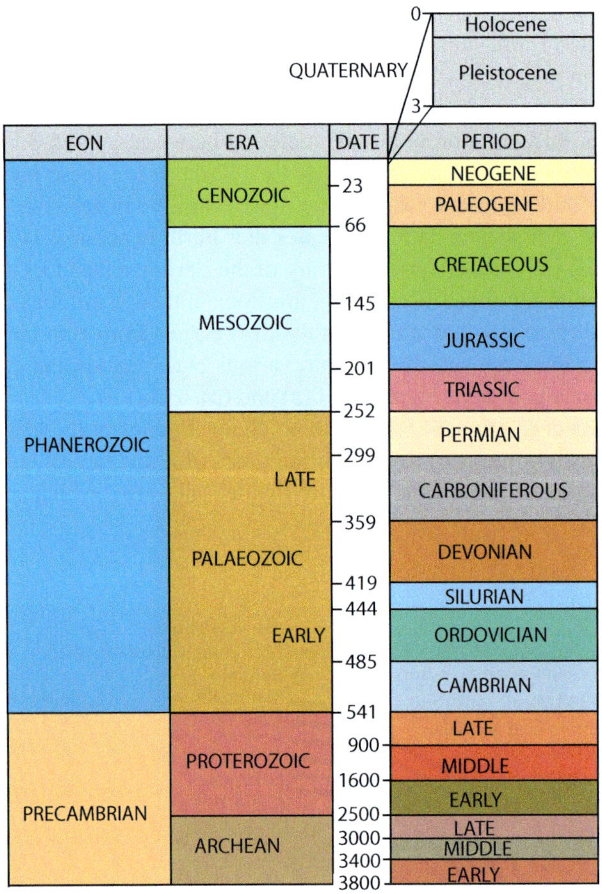

Figure 2.10(b) Geological time chart showing the relative order and time intervals represented by the different geological eons, eras and periods.

into **periods** (see Fig. 2.10(b)). We will sometimes use the geological periods listed in the right-hand column in Figure 2.10(b) when referring to the rock types in this book. Note that the Quaternary is the last period of the Cenozoic era and contains two epochs, the Pleistocene, which was a period of extensive glaciation and the Holocene, the current inter-glacial epoch which started about 12,000 years ago.

The origins of the period names reflect different modes of derivation: for example, the Cambrian is named after Cambria, the Latinised form of Cymru (Welsh for Wales), and the Silurian is named after the Silures, a Celtic tribe from South Wales. The intervening Ordovician rocks are named after the Celtic tribe Ordovicies who inhabited North Wales. Carboniferous means 'coal bearing' from the Latin *carbo* (coal) and *fero* (I bear, I carry). The Permian is named after the city of Perm near the Ural Mountains and the name Cretaceous is derived from the Latin *creta*, meaning chalk.

2.7 THE DISTRIBUTION OF ROCK IN THE BRITISH ISLES AND ITS SIGNIFICANCE FOR BUILDING STONES

At the beginning of this chapter, we included the 18th century quotation from William Cowper: "The Earth was made so various, that the mind of desultory man, studious of change and pleased with novelty, might be indulged." The geological map of Britain shown in Figure 2.11 certainly underlines the thrust of this quotation by illustrating the distribution of rocks of different types and ages—from the oldest rocks which crop out in the Outer Hebrides in Scotland and were formed in Pre-Cambrian times (the Lewisian gneiss is around 3000 million years old, i.e., two thirds the age of the Earth, Fig. 2.10(a)), to the relatively young rocks of the Tertiary period which crop out in the SE of England, e.g., the region around

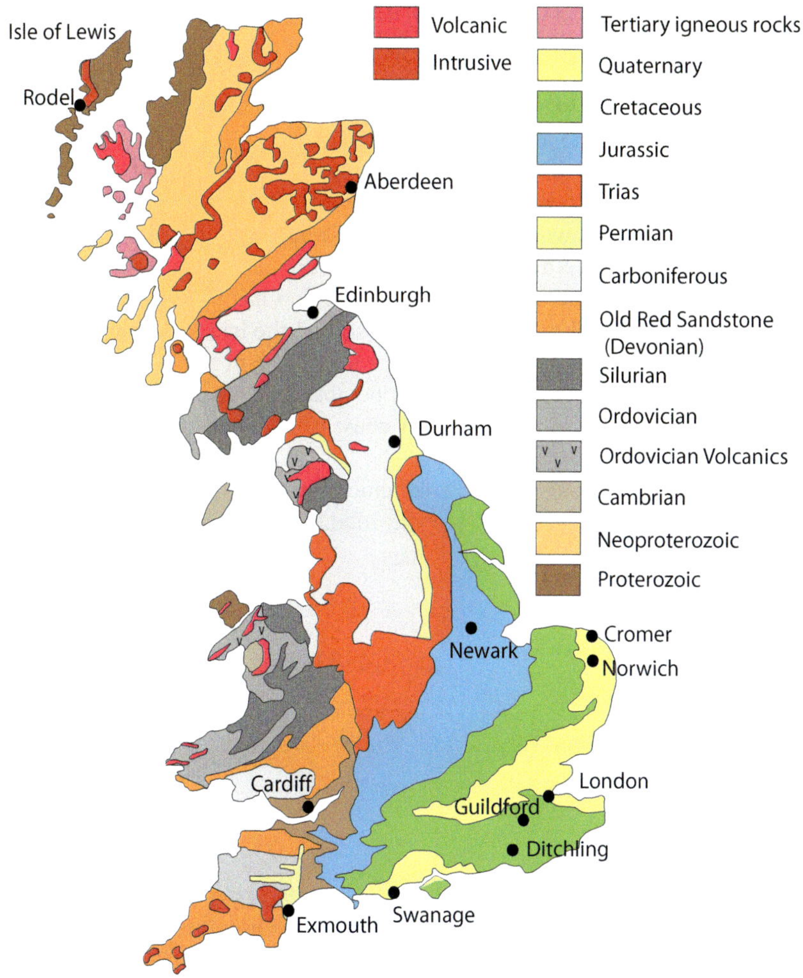

Figure 2.11 A simplified map of the geology of Britain. (The highlighted towns are referred to in various parts of the text.)

London, and which represent the Eocene, the London Clay having been deposited around 50 million years ago, close to the 12 o'clock position in Figure 2.10(a).

A particularly striking feature about the distribution of the different rocks in Britain as shown in Figure 2.11 is their tendency to outcrop in NE–SW trending strips with the youngest rocks occurring in the South-East and the oldest in the North-West. This outcrop pattern is the result of the tilting of the UK about a NE–SW trending axis as a result of the on-going collision between the African and Eurasian tectonic plates (Fig. 2.6) which began in the Late Cretaceous some 70 million years ago and is currently continuing to generate the Alps in southern Europe.

Rocks are laid down 'oldest first–youngest last' and tend to form in what is termed a 'layer-cake stratigraphy'. This process can be interrupted by the formation of a mountain belt when two plates collide locally deforming or tilting the strata which is then uplifted and eroded before the process of sedimentation takes over once more, Figure 2.12. A NW–SE vertical section drawn from London to the Isle of Lewis in Figure 2.11 would show progressively older rocks being exposed at the surface as one travels NW. Drawing (c) in Figure 2.12 is a schematic representation of this effect where the beds become older as one moves across the unconformity from right to left.

An E–W section along the south coast of England, although not at right angles to the axis of tilting, also illustrates this effect, see Figure 2.13 which shows a schematic cross-section between the towns of Exmouth and Swanage. This cross-section is similar to that shown in Figure 2.12(d), except that (since the deposition of the Upper Greensand and the Chalk above the unconformity) the region has been subjected to a further episode of tilting linked to the Alpine collision so that the beds above the unconformity are now themselves tilted. Uplift linked to this last phase of tilting has resulted in erosion and the current level of erosion that can be observed when travelling East to West from Swanage to Exmouth along the south coast of England is shown by the dashed black line in Figure 2.13.

During the movement of the Earth's tectonic plates, subduction can occur, i.e., one plate can be forced beneath another, see the subduction zones at the convergent plate

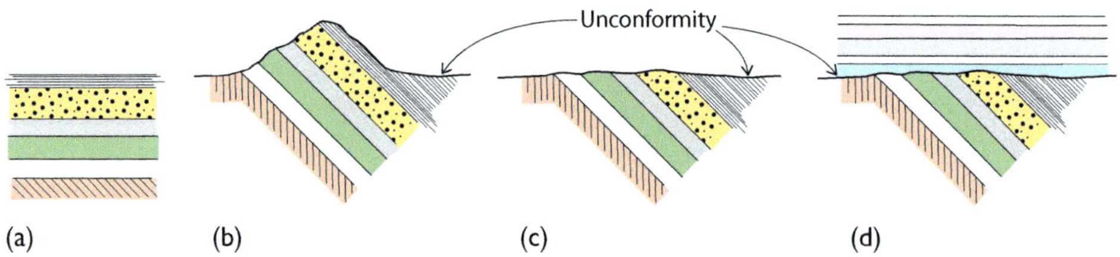

Figure 2.12 The impact of an orogenic event (tectonic plate collision) on a rock succession. (a) 'Layer-cake' stratigraphy develops as progressively younger beds are laid down on older beds. (b) The interruption of this process by an orogenic event which tilts and uplifts the strata. (c) The erosion of the uplifted block prior to (d) the deposition of younger rocks, **unconformably** on the older rocks.

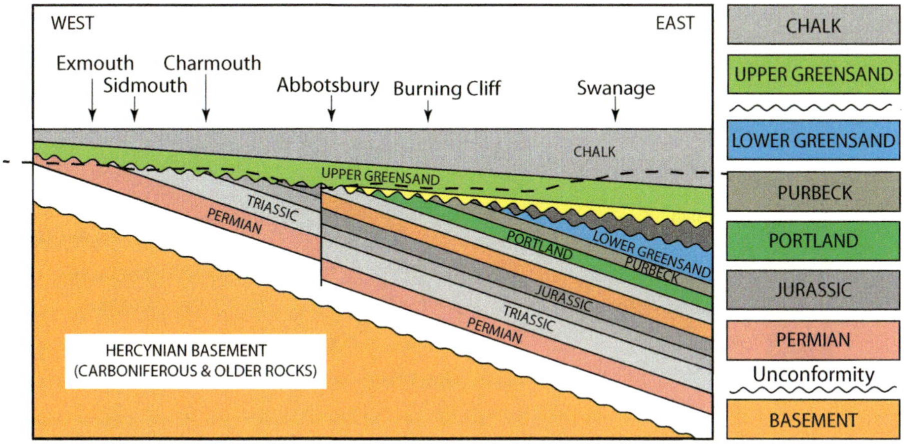

Figure 2.13 A schematic cross-section between Swanage and Exmouth (see Fig. 2.11) in England. The dashed black line shows the approximate position of the present level of erosion.

margins in Figure 2.6. The subducting plate is heated up and begins to melt, producing large volumes of magma which can be injected into the overlying rocks where it either cools slowly at depths as granite masses or cools quickly as a result of extrusion at the Earth's surface as a volcano. These granite masses and volcanoes form along the line of the mountain belt, i.e., parallel to the boundary between the two converging plates. Two such chains of granites, linked to ancient collisions between plates, can be seen in Figure 2.11. One runs NE–SW from John O'Groats SW into Northern Ireland. This is linked to the closing of an ancient ocean, the Iapetus ocean, which occurred in the Palaeozoic Era during the Cambrian and Ordovician periods ~500–450 million years ago, and which produced the Caledonian mountain belt.

The other belt is linked to the formation of an ~E–W trending mountain belt, the Hercynian belt, which formed at the end of the Carboniferous period, some 300 million years ago, as a result of the colliding and welding of two major plates, Laurasia in the north and Gondwanaland in the south, to form a single massive continent named Pangea. Some of the granites linked to this collision are exposed on the Cornubian peninsula (which is the South-West peninsula of Britain) and includes the Dartmoor, Bodmin, St. Austell, Carnmenellis, Land's End and Scilly Isles granites, Figures 2.11 and 2.14. This chain of granites is thought to link to a single massive body of granite at depths (a batholith) which underlies Cornwall and is, to some extent, responsible for the existence of the Cornubian peninsula. The dashed lines in Figure 2.14 are two important ancient NW–SE trending faults formed during the Late Carboniferous collision. Like the NW–SE faults shown in Figure 2.9 which were formed at the same time, these Cornish faults have also been reactivated on numerous occasions by the later Alpine collision.

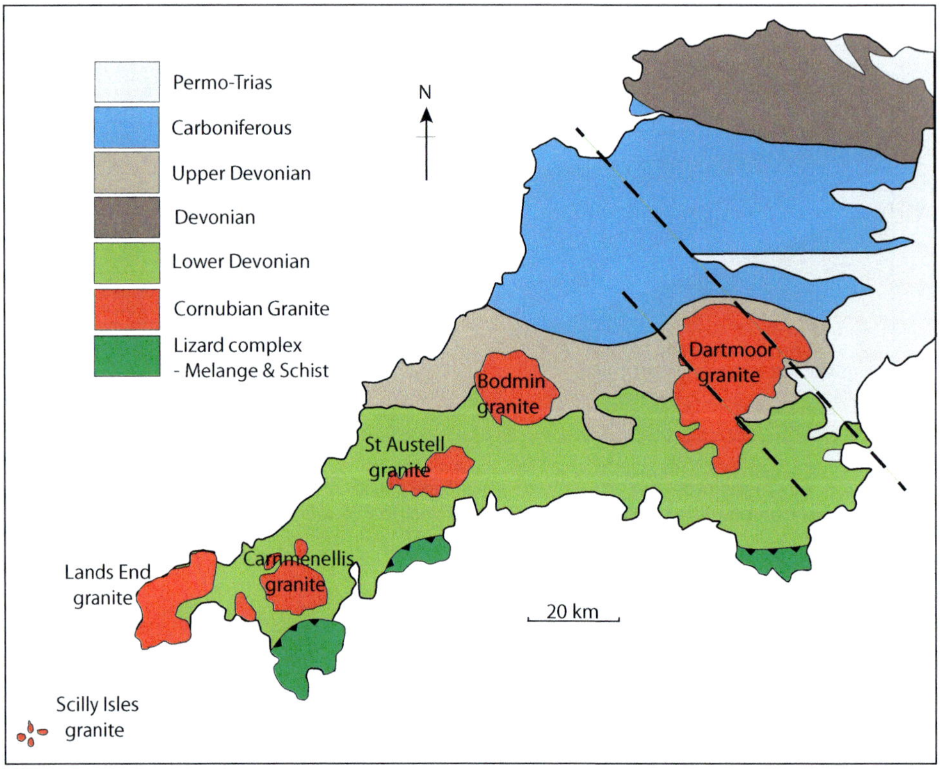

Figure 2.14 The granite exposures along the Cornish peninsula, South-West Britain.

2.8 THE NATIONAL STONE CENTRE AND THE BGS GEOLOGICAL WALK

There are two centres in England where British stone specimens from the different geological ages are available for public viewing. One is the National Stone Centre near Matlock in Derbyshire; the other is at the British Geological Survey's Headquarters at Keyworth, Nottingham, both in England.

The **National Stone Centre** is located at Porter Lane, Wirksworth, Derbyshire, England DE4 4LS. It is a Site of Special Scientific Interest with a half-mile GeoTrail, a Building Britain Exhibition, and a display of 19 different sections of dry stone wall, each built in a local style with stone from different regions.

The 10 levels in the stone staircase in Figure 2.15 are comprised of stone from different geological periods (see Fig. 2.10(b)). Climbing the staircase, one can successively inspect stones from 10 geological periods: 1. Pre-Cambrian; 2. Ordovician (Borrowdale Volcanics); 3. Ordovician (Early Caradoc Series); 4. Silurian; 5. Carboniferous; 6. Middle Carboniferous (Namurian Series); 7. Permian; 8. Triassic; 9. Jurassic; 10. Palaeogene (Antrim Basalt Group).

The **British Geological Survey (BGS) 'Geological Walk'** is located at the headquarters of the BGS at Keyworth in Nottingham, England. The emphasis here is placed on the variety of stone paving slabs that are available from the different geological periods, but

Figure 2.15 The National Stone Centre in England.

a selection of rocks and boulders is also on display, e.g., in Figure 2.16. Note the patterns in the Carboniferous Yorkstone paving slabs in this photograph: these are not caused by the original sedimentation process but by the later circulation of fluids and are known as liesegang rings, as we discuss in detail in Section 7.3 in relation to the building stones of Durham Cathedral. Along the 130-m Geological Walk, a comprehensive selection of 40 different types of building stone and boulders from different geological periods is presented. The focal point in the approach to the BGS Reception building is a monolith of Lewisian gneiss—a rock that is the oldest in Britain and one that we highlight in the next section.

There are also two museum type collections of British building stone: one is the early 20th century John Watson collection in the Sedgewick Museum of Earth Sciences (Geology Common Room) in Cambridge; and the other is the Natural History Museum's collection, which is on permanent loan to the BGS at the Keyworth site. However, neither of these collections is easily accessible by the general public.

Figure 2.16 Part of the Geological Walk at the British Geological Survey (BGS) Headquarters in Keyworth, Nottingham. (Courtesy of G. M. L. Baxter)

2.9 EXAMPLES OF STONE BUILDINGS AND THEIR GEOLOGICAL LOCATION IN BRITAIN

In this section, we illustrate some of the buildings in Britain that have been built using local stone for their construction. The sequence of north to south photographs follows a path from the Isle of Lewis in the Outer Hebrides, through Aberdeen and Edinburgh, then Durham, Newark and London, to Guildford in the south, i.e., travelling from the North-West to the South East of Britain, from the older to the younger rocks. The locations of the photographs in Figures 2.17–2.38 are highlighted on the geological map in Figure 2.11.

2.9.1 Isle of Lewis, Scotland

The rock forming the Isle of Lewis in the extreme north-west of Scotland is of Pre-Cambrian age (see the location in Fig. 2.11) and is a granitic gneiss formed by extreme metamorphism. It is similar to rocks in Canada and Greenland, indicating that Scotland was once part of the same land mass, Laurasia before being separated by the opening of the Atlantic Ocean, Figure 2.6. The Lewisian gneisses (named after the island of Lewis) contain amphiboles and pyroxenes (see Fig. 2.4), minerals rich in iron and magnesium. These weather to produce, amongst other minerals, iron oxides—which impart the golden hue to these rocks so evident in Figure 2.17.

This granitic gneiss building stone was used in the construction of St. Clement's Church in Rodel, Isle of Harris in the Outer Hebrides, Figure 2.17. The simplicity of the architecture plus the harmonising colours of the individual building stones combine to make this a delightful example of 'vernacular architecture'.

The typical landscape of the Isle of Lewis in Scotland, shown in Figure 2.18, indicates not only the easy availability of rough building stone but also the low relief typical of an ancient mountain belt (the Caledonian mountains) whose original topography has been removed by erosion to expose high grade metamorphic rocks (the Lewisian gneiss) which were generated at considerable depths in the Earth's crust under conditions of high pressure and temperature. The detailed, hummocky topography with surface cobbles and boulders reflects the fact that the region has relatively recently experienced glaciation (i.e., during the last ice age some 12,000 years ago).

Although this book is about building stones and stone buildings, there are many related aspects, one of which is the fact that there was a proposal for a 'superquarry' at Rodel—intended to extract rock for use as aggregate for road core, i.e., 600 million tonnes of rock

Figure 2.17 The tower of St. Clement's Church built of Lewisian gneiss in Rodel, Isle of Harris in the Outer Hebrides, Scotland, viewed from the south, left photo and the west, right photo in the afternoon sunshine.

Figure 2.18 Typical landscape on the Isle of Lewis in Scotland.

over 60 years. As a result of its geological origin, the local rock is extremely hard and resistant to deterioration and so is ideal when crushed for use as an aggregate. At the end of the proposed quarrying, a 1 km wide by 2 km long 'crater' with a depth of 180 m below sea level would have been created. But after 13 years of preparatory work by the promoters, together with associated discussion, the Inquiry (the longest in Scottish planning history) eventually found against the development. Note that quarrying/mining for building stones has a different emphasis when compared to quarrying for aggregate. We discuss the associated geological and rock mechanics aspects of the various quarrying procedures in Section 4.2.

2.9.2 Aberdeen

Travelling South-East across Scotland from the Isle of Lewis to the city of Aberdeen on the North Sea coast, we note that Aberdeen is known as 'the Granite City' because many of its major buildings are built using granite. This stone has a high strength, is resistant to water ingress, and can take a good polish—making it ideal as a building stone, both structurally and decoratively. For these reasons, granites from Aberdeenshire and from Cornwall have played a large part in British buildings. For example, the silver-grey granite from Kemnay (a village 16 miles, 26 km, west of Aberdeen) was used for the Forth Railway Bridge, the War Memorial in Aberdeen, Figure 3.11 and the Scottish Parliament building, plus the setts (cobbles) for British roads.

The following series of photographs in Figures 2.19–2.22 illustrates the use of granite in Aberdeen. Figure 2.19 shows the typical style of architecture via the Royal Bank of Scotland building; Figure 2.20 is of part of the granitic 'perpendicular gothic' style Marishal College building which houses mainly the Aberdeen City Council; and the texture of the Kemnay granite can be seen in the carving in Figure 2.21.

Earlier, with reference to Figure 2.4, we noted that granite, an igneous rock, is composed of feldspars, quartz and mica, and that large crystals of feldspar can sometimes be seen in granites (Fig. 2.3). This is again illustrated in Figure 2.22 showing studded pedestrian pavement slabs containing white feldspar (plagioclase) phenocrysts. Such tactile warning

Figure 2.19 The granitic façade with its Corinthian style pillar decoration of the Royal Bank of Scotland in Aberdeen.

paving surfaces are commonly seen in British towns to assist the visually impaired, but are usually made of synthetic materials or existing slabs retrofitted with stainless steel studs. In Figure 2.22, note that there can be portions of the same white phenocryst in both the stud and the background stone, e.g., in the third row down; this indicates that the studs have been created by a milling process, rather than being fixed on separately. In the case of such round integral studs, the stone milling is more complex than that for square blisters and so it was probably completed using a hollow diamond drill bit.

Traditionally, the granite for the buildings in Aberdeen was sourced locally, especially from the Rubislaw Quarry, but this closed in 1971 and now some of the granite for new projects is being sourced from overseas, e.g., from China for the Marischal Square development scheme, which is directly opposite the Marischal College building and was completed in 2018.

Figure 2.20 Marishal College in Aberdeen, said to be the second largest granite building in the world (i.e., after El Escorial, the historical residence of the King of Spain, 28 miles NW of Madrid).

Figure 2.21 Kemnay granite carving at the entrance to the Marishal College, Aberdeen. (Scale indicated by the pen in the lower right-hand corner.)

Figure 2.22 Pavement studs in Union Street, Aberdeen. A grey muscovite-biotite granite with white plagioclase phenocrysts clearly visible.

2.9.3 Edinburgh

Moving south, but still in Scotland, the buildings in Edinburgh are constructed with granite and sandstone, the red sandstone giving Edinburgh its red stone appearance. The next set of photographs illustrates Edinburgh's characteristic and charismatic sandstone buildings built with New Red Sandstone from the Permian and Triassic periods, much of it quarried locally, Figures 2.23–2.26.

The use of the term 'ashlar' in the caption to Figure 2.25 refers to the masonry of the pillar bases and the wall pilasters (on the left and right sides of the photograph) that consist of stone blocks which have been finely worked to produce flat surfaces, right-angled corners and thin joints.

2.9.4 Durham

Travelling south from Edinburgh lies the city of Durham which is also renowned for its sandstone buildings including its magnificent cathedral. In Figure 2.27, we show its majestic towers which were built in the period 1093–1133 AD and, in Figure 2.28, the problem of the Cathedral building stones which have been subjected to significant weathering.

The problem of dealing with weathered stone is intractable. What does one do when a building stone decays until it no longer retains its structural integrity, let with 'alone' its attractive

Figure 2.23 The former Armenian Restaurant and Community Cultural Centre (built in 1896 and previously the old Police Station) at 55 Abbeyhill, near the Palace of Holyroodhouse and the Scottish Parliament Building. Note the traditional Edinburgh architecture with the finialled, conical roofed, fishscale-slated, 'candlesnuffer' type turret roof ornaments, plus the parapet.

Figure 2.24 Pilasters on the south west wall of the Scottish National Gallery, Edinburgh. The bedding in the sandstone is evident.

Figure 2.25 Edinburgh University College of Art, main building built in 1906–9, red sandstone ashlar.

Figure 2.26 The Corbusian forms of the National Museum of Scotland Extension in Edinburgh, post-modern style, Permian sandstone, Clashach Quarry. Many of the interior and exterior sandstone blocks exhibit dominant markings which are not related to the sedimentation process but to later chemical reactions forming 'liesegang rings'; these have been mentioned previously in this chapter and are discussed in detail in Section 7.3 with reference to their occurrence in the Durham Cathedral stones and their effect on the weathering process.

Figure 2.27 Durham Cathedral viewed from the River Wear valley.

Figure 2.28 Replacement building stones in an exterior wall of Durham Cathedral. Note the characteristic weathered texture of the original sandstone building stones.

appearance? The solution illustrated on the outside of Durham Cathedral in Figure 2.28 is to replace a deteriorated building stone with a new building stone. Whilst this may enhance the strength of the wall, it does nothing for the overall attractiveness of the surrounding weathered building stones. We discuss the specific case of the Durham Cathedral sandstone problem in Section 7.3 and the general subject of the deterioration of building stones in Chapter 8.

2.9.5 Newark

Continuing our journey south through Britain we arrive at the town of Newark on the River Trent. Here we find examples of Ancaster stone, a yellowish Middle Jurassic oolitic limestone, from Lincolnshire to the east of Newark. In Chapter 3, we explain how to recognise many of the different geological types of building stone and Ancaster stone is one of the easily recognisable types because of its irregular banded appearance—as seen in Figure 2.29.

Ancaster stone and other types of building stone are also visible in the ruins of Newark Castle, Figure 2.30. It is said that King John died in Newark Castle from a surfeit of peaches.

2.9.6 London

London, located on the Thames river in the south of England, dates back to Roman times (as Londinium) and is thought to have had a population of around 60,000 at that time. Much later, in the 1800s, it had grown to around one million inhabitants and to more than five million in 1900, when it was the largest city in the world and the seat of the British Empire. Directly associated with the city's population growth and prestige was a marked increase in

Figure 2.29 Images of the characteristic texture of Ancaster stone at St. Mary Magdalene church in Newark, England.

Figure 2.30 The picturesque ruins of Newark Castle.

not only house building but also prestige buildings. The Cretaceous Kentish ragstone (which we highlighted at the beginning of the book in the Figure 1.1 photograph of the Jewel Tower in Westminster) was used widely in earlier days and Portland stone from the Jurassic period was the stone *de rigueur* for the prestige buildings, such as St. Paul's Cathedral, the British Museum, the Bank of England, Figures 2.31(a) and 2.31(b), and the National Gallery.

Transport of building stones was a problem in the earlier days but London's location on the Thames river with access from the sea facilitated water transport for both the Kentish ragstone and especially the Portland stone from the south coast. The current mining of Portland stone is illustrated and described in Section 4.3.2. Nowadays, building stones and particularly decorative stones from all over the world can be seen on and in London's buildings, e.g., the spectacular marble displays in the Roman Catholic Cathedral on Victoria Street. An excellent source of descriptive material relating to the building and decorative stones of London can be found in the Further Reading section of the excellent http://londonpavementgeology.co.uk website.

2.9.7 Guildford

The town of Guildford is located south-west of London in the county of Surrey and, like London, does not have a ready supply of high strength local building stones apart from flint. In Figure 2.32, a wall at the base of Guildford Castle is shown with three layers of building stone—from top to bottom: flint, sandstone and chalk.

Brick, which can be considered as an artificial building stone (see Section 3.9.3), has been widely used in Guildford, not least in the building of the impressive 20th century Guildford Cathedral (Fig. 2.33). After the Second World War in the 1950s, a 'Buy a Brick' campaign was established in order to fund continuing construction of the Cathedral. Of the

Figure 2.31(a) Façade of the Bank of England in London constructed with Portland stone.

Figure 2.31(b) Corinthian style pillars, Portland stone, Bank of England, London.

Figure 2.32 Wall at the base of Guildford Castle built with, from top to bottom, flint, sandstone and chalk.

Figure 2.33 Guildford Cathedral built with bricks.

Figure 2.34 One of the Guildford cathedral bricks signed by Queen Elizabeth II.

order of 200,000 bricks were bought and the buyers were entitled to sign their name on the brick. A brick which has been signed by Queen Elizabeth II is highlighted in Figure 2.34.

In addition to the local bricks, other non-local building stones have been used: piers of the nave arches are faced with Middle Jurassic Doulting limestone from Somerset and the nave floor is made of Travertine marble from Italy. This type of modern cathedral is not expected to appeal to everyone's architectural taste but it is well worth a visit to experience the site at the top of Stag Hill, the lightness and simplicity of the design, and the sheer effort in the one-at-a-time funding and consequent manufacture of so many bricks.

2.9.8 South-east England

Finally, in these brief sketches of how the local geology affects the use of different types of building stone, we consider the south-east of England. It can be seen in the earlier Figure 2.11 geological map of the British Isles that the surface of this area is composed of Cretaceous and younger rocks. There is a strong correlation between the age of a rock and its strength and durability as a building stone, the older the stronger (see Fig. 5.2), with the consequence that there is a paucity of suitable highly durable building stones in the south-east of England. However, the exception to this general principle is the existence of flint in the chalk strata. Flint is composed of silica (i.e., quartz), SiO_2, which is both strong and durable but, unfortunately and unlike other building stone material, it does not occur in convenient geometrical sizes or shapes, Figure 2.35. Although relatively plentiful in the Middle and Upper chalk strata, the resistant flint is not present in the Lower Chalk (much to the relief of the Channel Tunnel boring machines when they excavated the tunnels between England and France in the Lower Chalk stratum!).

Flint has been worked in south-east England since around 2000 BC and particularly at Grime's Graves in Norfolk, where there are between 700 and 800 shafts and shallow pits over an area of 34 acres, the flint being used mainly for making tools. Since then, the essentially impermeable flint has been used extensively on the exterior walls of buildings, Figure 2.36 showing the exterior of the entrance to the Guildhall in Norwich, the county

Figure 2.35 Exterior of St. Margaret's Church, Ditchling, SE England, showing the 'as found' flint shapes and the long-term durability of flint as compared to the local shaped stone.

Figure 2.36 The flint and stone façade of the Guildhall, Gaol Hill, Norwich, Norfolk, England.

town of Norfolk, and Figure 2.37 illustrating the shaped, square, flint blocks which form the exterior wall. Figure 2.38 highlights the flint and terracotta east elevation of the Methodist Church, West Street, Cromer in Norfolk, England.

Finally, we note two creative examples of the use of flint. In the mortar of old stone buildings, flint chips can often be found, a procedure known as 'galletting' (*galet* is French

Figure 2.37 The flint blocks ('knapped' to square shapes from the irregular shapes in which the flint naturally occurs) forming the façade of The Guildhall, Gaol Hill, Norwich, Norfolk, England.

Figure 2.38 Flint and terracotta east elevation of the Methodist Church, West Street, Cromer, Norfolk, England.

for pebble) which is used to save money—because the flint chips are cheaper than the mortar. However, in Figure 2.39, flint chips have been used to cover completely the exterior front wall of a house. In the second example, Figure 2.40, flint cobbles have been used creatively to fill wire cages (known as gabions) forming part of the bus station in the town of Cromer in Norfolk, England. Notice that the cobbles are rounded—probably as a result of being

Figure 2.39 Unusual front wall decoration/protection, Flint House, East Runton, Norfolk, England.

Figure 2.40 Flint gabions, Bus Station, Cromer, Norfolk, England.

sourced from the nearby beach. See also the gabions on the dedication page at the beginning of the book.

* * * * *

This concludes Chapter 2 on the geological origin of building stones. In the next chapter and based on the knowledge in this chapter, we discuss the main British building stone types and how to recognise them.

Chapter 3

Recognising the different types of building stone

This chapter is dedicated to B.C.G. Shore, author of the 1957 book Stones of Britain.

3.1 INTRODUCTION

Our main motivation in writing this book is to assist readers in identifying different types of building stone and the associated architecture. As we walk in the streets of our cities, towns and villages, life is more interesting if we can identify the more common building stones and appreciate the architecture that they create. The frequent use of any one particular type of locally obtained stone in a village or town leads to a vernacular architecture and hence to that town's character. However, building stones have often been transported across the country, to support a particular architectural theme, such as the frequent use of the white Portland stone for the important buildings in London and other cities; alternatively, a mixture of building stones from different sources may have been used. In either case, it is satisfying to understand the built environment via recognition of the different types of building stone. For this reason, we now follow the explanations of the geology in Chapter 2 with descriptions and illustrations of granites, volcanic stones, limestones, sandstones, flint, metamorphic stones, breccias and conglomerates, plus a range of different types of artificial stone—highlighting those features that enable a particular building stone to be recognised. Chapter 4 follows describing the life of a building stone, and then the elements and types of stone buildings are outlined in Chapter 5 and the different styles of architecture in Chapter 6.

3.2 GRANITES

Granite is a hard, granular, crystalline, igneous rock consisting mainly of quartz, feldspar, and mica.

3.2.1 The use of granite

Granites as building stones have important roles not only because of their decorative appeal, both when 'rough hewn' and when polished, but also because of their resistance to water penetration. Figure 3.1(a) shows Tower Bridge across the river Thames in London and Figure 3.1(b) illustrates the crystalline nature of the Cornish granite (from the Cheesewring quarries situated on the eastern flank of Bodmin moor) which was used to clad and protect the bridge structure. Figure 3.2 demonstrates the use of granite for a protective sea wall in

(a) (b)

Figure 3.1 (a) Tower Bridge across the river Thames. (b) The microstructure of the cladding granite used to protect and decorate the structure. Note that the white crystals are feldspar, the black crystals are mica and the grey ones are quartz. The tight interlocking of these crystals causes the granite to be essentially impermeable.

Figure 3.2 Protective granite sea wall at Wakkanai, Hokkaido, Japan.

Japan and Figure 3.3 illustrates two examples where granite has been used at the base of pillars to avoid water damage caused by rainwater collecting at the base of the pillars.

Granite has major structural advantages as a building stone: it has a high strength, it is resistant to water penetration and is extremely durable. Its use in the cladding of Tower Bridge, Figure 3.1(a), takes particular advantage of the last two of these qualities. Similarly, granite's properties are put to good use in the sea wall at Wakkanai in Hokkaido Island in Japan, as indicated in Figure 3.2. However, a historical disadvantage of granite in Victorian times and in the British context was that it occurs mainly in Scotland and Cornwall, thus necessitating significant transport across the country from the quarry to the building structure. Nowadays, this disadvantage in using 'homegrown' granite is mitigated by the improved transport infrastructure.

The vulnerability of building stones to fluid ingress is characterised by a property known as the *hydraulic conductivity* when water is the fluid, and by the *permeability* in the case of

(a) (b)

Figure 3.3 Use of two types of granite at the base of pillars to avoid water damage to the sedimentary pillar stone above.

other fluids. Assuming that there are no fractures in the stone, water will travel through the small pores in the stone—so the susceptibility of a particular building stone to water ingress will depend on the geometry of the microstructure, i.e., whether there are open pores and whether these pores are connected to form a hydraulic pathway. Thus, because of their crystalline microstructure (i.e., that of an interlocking mesh of crystals between which there are no significant pores), granites are much less susceptible to water damage than sedimentary stones such as sandstone and that is why (Fig. 3.3) it is always good practice to use granite as a foundation stone for external walls and pillars constructed using sedimentary stones, such as Portland stone (which is a limestone), or the building will be susceptible to the type of water damage shown in Figure 3.4.

Granites are known as *holocrystalline* rocks, the term describing a crystalline igneous rock consisting completely of crystals. The usual parameter characterising the mechanical strength of a stone is the compressive strength, i.e., the amount of stress (force per unit area) that a sample of the stone can sustain before it breaks. Because of the mechanically

Figure 3.4 Water damage to a sedimentary stone at the base of a decorative pillar.

integrated microstructure of the granites, they have the highest strengths of all the different building stones and this is why they give such good service, not only in the context of the normal weather conditions but also in the more aggressive conditions such as experienced by lighthouses and sea defenses, illustrated in Figure 3.2, plus bridge foundations and dam structures. Note that a case study of the use of Ross of Mull granite in the construction of the dramatic Stevenson Skerryvore lighthouse building off the west coast of Scotland is presented in Section 5.2.2.

In the context of granite decorative building stones, although the general texture consists of irregular and medium to coarse-sized grains, the feldspar can occur in the form of conspicuous tabular crystals of the order of 25 mm in length, known as phenocrysts (Figs. 2.3 and 3.5). This indicates that the feldspars were the first minerals to crystallise out from the magma and also that cooling occurred slowly at depths (the slower an igneous rock cools, the larger are the crystals). Feldspar is the most common mineral on Earth, constituting approximately 60% of the crust; it forms directly from cooling magma and is a major component of granite and most of the other igneous rocks; two common feldspars that occur in granite are orthoclase (a potassium rich feldspar) which is often pink, and plagioclase (a calcium and sodium rich feldspar) which is often white. On an average, granite contains about 60% feldspar and 25% quartz (SiO_2), a glassy, colourless mineral—in contrast to the micas which occur in both dark and light forms, the dark form being biotite and the light form being muscovite. Several decorative granite surfaces are illustrated in the next section.

Figure 3.5 Polished granite surface containing a white feldspar phenocryst in the Merrivale granite from Dartmoor—on the exterior of the New Scotland Yard building in London.

As illustrated in Chapter 2, the granitic outcrops in Britain occur mainly in Cornwall and Devon, and north of the Highland Boundary Fault in Scotland, but there are other important outcrops in England, such as the Shap granite in Yorkshire and the granodiorite at Mountsorrel in Leicestershire, noting that this latter rock is used for aggregate purposes. The granites generally form after major orogenic (mountain building) events and the two events that dominate the granites of Britain are the Late Carboniferous Hercynian orogeny which gave rise to the granite spine of the Cornubian peninsular and the earlier, Caledonian (Silurian) orogeny which accounts for the granites of the Lake District and Scottish Highlands.

3.2.2 Decorative granite surfaces

Polished granite surfaces are widely used as decorative surfaces for buildings, both internally and externally. They exhibit a variety of attractive textures and, for external use, they have the added attractions of being resistant to both the weather and mechanical damage. A variety of polished granite surfaces are shown in Figure 3.6.

As can be seen from Figure 3.6, the textures and colours of granites are varied. Their colour is generally controlled by the colour of the feldspar, the main mineral constituent; feldspars are aluminium silicates and their colour is determined by the alkali metals they contain. Orthoclase is a pink feldspar and contains potassium; plagioclase is a white to grey feldspar and contains sodium and calcium. The Peterhead granite in Figure 3.6 is dominated by orthoclase which gives it its characteristic pink colour, and the example illustrated contains a small black inclusion of the rock into which the granite was intruded. Depending on

Peterhead granite, Scotland, with a black xenolith (foreign inclusion).

Bodmin Moor granite, Cornwall.

Kemnay granite, Aberdeen, Scotland.

Rapakivi granite, Finland.

Figure 3.6 Four examples of decorative granite surfaces.

their composition and the time they spent in the magma before it cooled and solidified, these xenoliths (literally foreign rocks) may become completely assimilated by the magma or may remain intact as in the example illustrated here.

The Bodmin granite shown in Figure 3.6 is dominated by plagioclase, i.e., white feldspars. The tabular shape of these crystals is apparent and, as noted above, reflects the fact that these were the first minerals to crystallise from the magma. When conspicuous crystals are larger than the grains of the rock groundmass, they are termed phenocrysts. Minerals such as quartz that crystallise out later, have to crystallise in the gaps within the crystal mix of feldspars and, as a result, the quartz crystals are smaller and less well formed. The Kemnay granite shown in Figure 3.6 is also dominated by plagioclase feldspars. It is finer grained than the Bodmin Moor granite, indicating that it cooled more rapidly. The Rapakivi granite which is also shown in Figure 3.6 has an easily recognisable texture: the feldspars form large, ovoid phenocrysts 10 to 50 mm in diameter set in a finer groundmass. The feldspars are composite, i.e., made up of two types: the centres are pink or brown orthoclase which is entirely mantled by grey plagioclase. The word 'Rapakivi' in Finnish means 'weathered rock'.

In Figure 3.7, we illustrate another example of an ornamental granite, this time from Sweden, and in Figure 3.8 the decorative nature of a syenite is shown. Figure 3.9 illustrates a

Figure 3.7 Unweathered Swedish Imperial Red granite with large plum red feldspar crystals.

Figure 3.8 An example of a syenite—which is similar to granite but contains less quartz.

Figure 3.9 Weathering of Shap granite (from Cumbria, England).

polished surface of Shap granite; this well-known, highly prized and decorative granite contains large phenocrysts of pink (i.e., orthoclase) feldspar set in a finer grained matrix made up of orthoclase, plagioclase (white), quartz (glassy) and biotite mica (black). Despite the resistance of granite to weathering as indicated by the many monuments that have survived in almost pristine form for centuries, under certain conditions the feldspars can deteriorate. The onset of the breakdown of feldspar can be seen in Figure 3.9. The best known example of this is the formation of china clay (kaolin) in Cornwall resulting from the breakdown of feldspars by the acidic fluids that pass through granite masses in the later stages of their cooling. These fluids do not affect the quartz which is the reason why amethysts can be found in the china clay pits. The aggressive acidic atmospheres of urban environments can also attack the feldspars.

3.2.3 Buildings built with granite and carved granite features

As we have noted, granite can be used both as a building stone and as a decorative stone. Buildings built with granite tend to be in areas where granite is abundant, such as Aberdeen (known as 'The Granite City') in Scotland, which was highlighted in Chapter 2. Given the tough adamantine nature of granite and hence its use as a resistant stone, both for structural building and decorative facings, one would expect that it is not the easiest material to shape and carve—and that is indeed the case—but it does have its advantage of a high resistance to weathering, and features carved in granite do have their own special appeal, as evidenced in Figures 3.10–3.15.

In Figure 3.15, two of Wrocław's bronze dwarfs use a granite ball to provide a physical illustration of Newton's third law—that every action has an equal and opposite reaction. These figurines first appeared in the streets of Wrocław in 2005 and spotting them provides an interesting tourist activity.

Figure 3.10 The granite Central Library in Aberdeen, Scotland.

Figure 3.11 Granite war memorial in Aberdeen, Scotland. The rugged granite lion was designed by Aberdeen sculptor William Macmillan with the work carried out by James Philip using silver-gray Kemnay granite, see Figures 2.21 and 3.6.

(a) The façade with its muted Corinthian style pillars.

(b) Detail of the granite used in its construction.

Figure 3.12 The Parliament building in Helsinki, Finland.

(a) South Silver Street, Aberdeen, Scotland.

(b) Railway Station, Helsinki, Finland.

Figure 3.13 Building decorations carved in granite.

Figure 3.14 Carved Art Deco granite decoration on a building in Helsinki, Finland.

Figure 3.15 Street decoration in Wrocław, Czech Republic.

Figure 3.16 Granite setts, outside St. Giles' Cathedral in the Royal Mile between Edinburgh Castle and the Palace of Holyroodhouse in Scotland, illustrating the advantage of granite's mechanical durability.

3.2.4 Further examples of the use of granite

In the following Figures, 3.16–3.18, four further examples of granite's properties are illustrated. In Figure 3.16, granite's high resistance to mechanical wear and tear is shown by its use as cobbles/setts in the refurbished road outside St. Giles' Cathedral in Edinburgh, Scotland.

(a) (b)

Figure 3.17 (a) Fracture through a large diameter granite drill core from the base of the Three Gorges Hydroelectric Dam site on the Yangtze River, China. (b) A long-lasting granite runestone, Sweden.

Figure 3.18 Spalling of a granite surface in the Gobi Desert, China, caused by exposure to the sun. (Image $\cong$ 0.5 m across)

In Figure 3.17(a), a sample of the mechanically strong granite below the Three Gorges Dam across the Yangtze River in China is shown and in Figure 3.17(b) the durability of granite during years of exposure to the Swedish weather is illustrated via the ancient rune-stone. Finally, in Figure 3.18, not even granite can resist the continual blistering heat of the Gobi Desert in China, where the repeated cycles of high temperature in the day and cooler temperatures at night have caused spalling of granite flakes at the exposed rock surface.

3.3 VOLCANIC STONES

A volcanic rock is an igneous rock resulting from volcanic action at or near the Earth's surface.

Although examples of volcanic rocks occur in all countries of the British Isles, they are not commonly used as building stones, but we include a brief discussion of them here for completeness. Significant outcrops occur in the Ordovician rocks of Snowdonia, North Wales, in the Ordovician rocks of the Lake District in England (the Borrowdale volcanics) and in the Tertiary rocks of the Inner Hebrides in Scotland. As noted in the above definition, volcanic rocks are igneous rocks that extrude onto the Earth's surface. They fall into two distinct groups, namely those that extrude as liquid magmas and produce lava flows and those that are ejected as dust, ash and larger fragments to produce what are termed pyroclastic deposits. These two types of volcanic rock can be used as building stones and, as illustrated in Figures 3.19–3.22, have characteristic features that can be used in their identification: in particular, Figure 3.22, the small holes known as *vesicles* which were formed by gas bubbles.

Given the violent origin of volcanic rock, we would expect it to be much more disturbed than granites or sandstones—and this is indeed the case as illustrated in Figures 3.19 and 3.20

Figure 3.19 Volcanic rock strata on the coast of Jeju Island, South Korea, illustrating the easily available, but awkwardly shaped, potential building stones.

Figure 3.20 Inhomogeneous and disturbed volcanic rock strata, Jeju Island, South Korea. The strata indicate a history of volcanic eruptions, with extremely fine particles forming some layers whilst other layers are interspersed with larger stones.

Figure 3.21 Dry stone wall built with volcanic stones, a vesicular basalt, Jeju Island, South Korea.

(a)

(b)

Figure 3.22 Characteristic holes (i.e., vesicles caused by hot gas) in volcanic lava, Jeju Island, South Korea. Note the variation in the size of the vesicles in image (a) and the gas pathway connections between vesicles in image (b).

from Jeju Island, off the southern coast of the Korean Peninsula. The island has the gentle topography of a shield volcano, so called because of its shape, a flattened dome, formed by successive eruptions of basaltic lava flows and the eruption of rhyolitic ash. Part of a basalt lava flow is shown in Figure 3.19; it became contorted as it flowed over an irregular surface and is now highly fractured. A pyroclastic rock from the same island is shown in Figure 3.20. These rocks are formed from the settling of ash, dust and coarser debris ejected from the volcano and can be deposited either in water or subaerially. They have the bedded nature of sedimentary rocks but are characterised by being very poorly sorted, i.e., they display a complete mixture of large and small particles. As can be seen, layers of fine ash (known as tuffs) are interbedded with coarser material. Some of the larger clasts (rock fragments) deformed the tuff layers onto which they fell, see the three clasts in the upper left-hand area of the photograph in Figure 3.20, showing that the tuffs were still unconsolidated sediments when the clasts were deposited. These clasts were subsequently covered by a layer of undisturbed ash, see the top of the photograph.

When studying and modelling the mechanics of rocks, the acronym DIANE is used as a cautionary reminder that rocks are likely to be Discontinuous, Inhomogeneous Anisotropic and Not Elastic. In other words, they are discontinuous if they contain fractures; they are inhomogeneous if they have different properties in different places; they are anisotropic if they have different properties in different directions; and they are not elastic if, when deformed, they do not return to their original shape. Needless to say, the volcanic rocks illustrated in Figures 3.19 and 3.20 are definitely DIANE materials!

The features shown in Figure 3.20, i.e., the poor sorting of the clasts and the deformation of the underlying tuff bands by the large clasts, enable pyroclastic flows to be easily recognised when used as building stones. Rocks formed by the extrusion of liquid magma also develop features that enable them to be instantly identified; the most common of these are vesicles, i.e., the cavities formed by gas bubbles trapped during solidification of the lava, as illustrated in Figures 3.21 and 3.22. The photograph in Figure 3.23(a) shows the façade of the Althing building, the National Parliament of Iceland in Reykjavík, which was built in

(a)

(b)

Figure 3.23 (a) The façade of the Althing building, the National Parliament of Iceland in Reykjavík, which was built in 1881 of local stone. (b) A close up of the vesicular volcanic stone used in the construction of the building. (Photographs courtesy of M. M. Hudson)

Figure 3.24 Ninth century monolithic Buddhist temple excavated in the Deccan basalts in India. (Photograph by the authors and courtesy of Imperial College Press, London)

1881 of local stone. The close-up of the volcanic stone used in the construction of this building, Figure 3.23(b) again demonstrates the characteristic vesicular nature of this volcanic type of stone. Note that Iceland lies on the boundary of the Eurasian and North American tectonic plates (Fig. 2.6 in Chapter 2) which are moving to the east and west respectively at a rate of ~25 mm/year. The volcanic rock that forms most of Iceland is the result of eruptions linked to this plate separation and these eruptions have given rise to the largest mountain chain on Earth—the Mid-Atlantic Ridge, stretching from the NE of Greenland to the South Atlantic. It is dominantly submarine but emerges occasionally to form the Mid-Atlantic Islands of which Iceland is the largest.

In our companion book, *Structural Geology and Rock Engineering*, we describe an interesting example of the use of volcanic rock in the 6th to the 10th centuries AD to carve *in situ* temples directly out of a basaltic rock mass at the Ellora World Heritage site in India, Figure 3.24. Internally, the temples are often decorated with carved motifs, as is illustrated in Figure 3.25. Note the primary horizontal layering of the basalt rock mass which is highlighted by the accumulation of vesicular horizons now infilled with zeolite (zeolites are a group of hydrated aluminium silicates which form where volcanic rocks and ash layers react with alkaline groundwater).

Needless to say, we are not familiar with molten magma erupting across the British countryside, although this has happened in the geological past. What is more in evidence now is magma that has been injected within rock strata in the past and cooled to become sills and dykes, such as the Whin Sill in Northumberland which was used by the Romans as a building stone for Hadrian's Wall. The wall is built on the topographic ridge generated by the erosion of the softer rocks on each side of the sill.

Figure 3.25 Monolithic elephant carving inside a temple at the Ellora site in India. (Photograph by the authors and courtesy of Imperial College Press, London)

3.4 LIMESTONES

A limestone is a sedimentary rock consisting of more than 50% calcium carbonate, $CaC0_3$.

3.4.1 Introduction

Following the previous sections in this chapter describing granites (which are widely used as a building stone for both structural and decorative purposes) and volcanic stones (which are not used extensively in Britain as building stones), we now come to limestones which form a large proportion of the building stones used in Britain—partly because of their advantageous mechanical and decorative properties and partly because of their availability. Two of the most well-known limestones are Portland limestone, which has been used throughout Britain as a structural building stone and is ubiquitous in central London buildings, and Purbeck 'marble', which is actually a limestone and has been extensively used in churches for decorative purposes. Illustrations of these two example limestones are provided in Figures 3.26 and 3.27.

(a) The Cenotaph, Whitehall, London. The UK's National War Memorial.

(b) Fossil remains in the Cenotaph stone, and as typically observed in Portland limestone.

Figure 3.26 Example of the use of Portland limestone—for London's Cenotaph, the empty tomb, designed by the architect Edwin Lutyens. The Cenotaph was unveiled on Armistice Day, 11 November 1920.

(a) Purbeck marble pillars, Temple Church, London.

(b) Close-up of a Purbeck marble pillar, composed almost entirely of fossilised *Viviparus* shells.

Figure 3.27 Example of the decorative use of Purbeck 'marble'.

Portland stone is an Upper Jurassic limestone and is a major structural and decorative building stone that has been used extensively in Britain, especially in London. The London Cenotaph is shown in Figure 3.26, the word 'Cenotaph' coming from the Ancient Greek for 'empty tomb'. In H. V. Morton's 1927 book *In Search of England*, he records a remark made by a foreman at the Portland stone quarry in the south of England: "I helped to choose some of the stones for the Cenotaph, the top ones with the wreath on them. We picked the purest white stone in the island." The Portland stone mine/quarry is highlighted in Chapter 4

The Jurassic Purbeck 'marble', Figure 3.27, found on the Isle of Purbeck near Swanage in England, was used as a decorative stone in earlier days because of the absence of any significant amount of true marbles occurring in Britain. However, Purbeck marble only occurs *in situ* in relatively thin strata and so columns, such as those in Figure 3.27(a), are made up of several drum-shaped sections. The appearance of the marble in Figure 3.27(b) is due to the fossilised shells of the small univalve freshwater snail *Viviparus*. The stone can hold a significant polish—which resulted in this limestone being colloquially termed a marble.

3.4.2 The distribution of limestones in Britain

In Section 3.2, we noted that the granites in Britain are focused in certain limited areas. By contrast, the limestones are much more widely distributed, not only geographically but also with respect to the times when they were formed, i.e., their geological periods. The authors have found that two books are particularly helpful when studying limestone building stones (see the References and Bibliography section for the full references):

- *Limestones: Their Origins, Distribution and Uses* by F.J. North;
- *The Building Limestones of the British Isles* by E. Leary.

The book by F.J. North is a 467 page *tour de force*. Part I concentrates on the nature, origin and varieties of limestones; Part II explains the distribution of limestone in the British Isles (see Table 3.1); Part III discusses limestone in relation to scenery, agriculture, water supply and industry. The building limestones report by E. Leary was generated by the Building Research Establishment, Department of the Environment, UK, and provides details of 30+ limestones—containing information about the source quarries, the petrology (the branch of geology dealing with the origin, occurrence and structure of rocks), reference buildings and the results of physical tests on the stones. Despite the fact that both books were published some years ago, 1930 and 1983 respectively, they provide an excellent foundation for understanding the nature and sources of building limestones in Britain. In particular, the reference buildings listings by E. Leary are helpful for those wishing to study a particular building stone in its full majesty at different locations. Note that there is also a matching E. Leary book on sandstones (see the references list at the end of the book). More detailed current information on specific limestone quarries is available through the Natural Stone Directory, e.g., No. 20, 2018–19 (QMJ Publishing Ltd., Nottingham NG1 5BS).

When a building limestone has no marked characteristics that are easily recognisable by eye, it is naturally difficult to identify. However, the following example limestones can be quickly recognised because of their special features: Portland stone, Ham Hill stone, Kentish ragstone and Ancaster stone. We now illustrate and describe each of these stones in the next sections.

Table 3.1 Limestone distribution in Great Britain, from F. J. North's 1930 book on limestones

Geological period	**Limestones**
Tertiary and Recent	Freshwater limestones of the Isle of Wight and recent travertines and tufas and stalagmites
Cretaceous	Chalk, Totternhoe stone, Beer stone, Kentish rag, Sussex marble
Jurassic (incl. Lias)	Many important oolitic limestones, for example Portland stone and Bath stone; and other building stones such as Douling stone, Dundry stone, Forest marble; limestones in the Lias are used for building and cement making
Triassic	Some conglomerates of limestone pebbles occur, and dolomitic grits, but no normal limestones
Permian	The Magnesian limestones of the eastern side of the Pennine range; building stones, dolomite, etc.
Carbonifereous	The lower part of the Carboniferous System consists largely of limestone—the Carboniferous limestone—which is used for building, roadstone, lime burning, chemical works, etc.
Devonian	Limestones used for building and ornamental purposes occur in the Middle Devonian of Devonshire; calcareous beds occur in the Old Red sandstone
Silurian	Limestones occur at various horizons in the Silurian System: they include the Wenlock limestone, Woolhope limestone and Aymestry limestone—used mainly for building, lime burning and metallurgical purposes
Ordovician	Limestones are not important except in the upper part of the Formation: Bala limestone in North Wales, Coniston limestone in the Lake District, Shoalshook limestone in South Wales
Cambrian	The Durness limestones and Dolomites of the north-west of Scotland are, in part, of Cambrian age, but Cambrian limestones of commercial importance are not abundant in England
Pre-Cambrian	Few limestones occur in the Pre-Cambrian rocks of this country; most of them are now crystalline marbles and few are of economic importance

3.4.3 Portland stone

We start with Portland stone, an oolitic limestone of Jurassic age, which is the most famous of all the southern British building stones, because it has been used to construct many prestigious buildings from the 17th century onwards, as stimulated by Inigo Jones, architect and surveyor to James I, who used it for the Banqueting Hall in Whitehall. Following the Great Fire of London in 1666, there was a major building programme with Sir Christopher Wren using the stone for his many churches. In fact, and as noted by North (1930), when Wren was overseeing the construction of St. Paul's Cathedral in London, the Portland Quarries were, by the king's command, worked exclusively to supply stone for the Cathedral. Extraction of stone at the current Portland stone quarry/mine is described in Section 4.3.2.

Some of the other impressive buildings in London constructed using Portland stone are the Bank of England (Fig. 3.28), British Museum, National Gallery, and Somerset House (Fig. 3.29). Additionally, numerous buildings in other towns in Britain have used Portland stone, e.g., the Fitzwilliam Museum in Cambridge and the extensive suite of buildings in Cardiff's civic centre.

Figure 3.28 Bank of England façade, Threadneedle Street, London.

Figure 3.29 Somerset House, London.

Many of the Portland stone buildings in Britain were constructed decades, if not hundreds of years ago, and hence have been subjected for some time to the ravages of the rain and frost, plus the smoke from earlier industrial and domestic coal fires. The associated weathering has tended to highlight the remains of contained fossils and hence enables Portland stone to be easily recognised following a few moments study, Figure 3.30. The most easily identifiable fossils are the oyster shells, as illustrated in Figure 3.31.

Figure 3.30 Weathered Portland stone with fossil remains, exterior wall of the Bank of England, London.

Figure 3.31 Fossilised oyster shell in Portland stone, Old Post Office, Howardsgate, Welwyn Garden City, England.

Figure 3.32 The Whitbed stratum of the Portland stone, Aldwych, London.

In the case of the Whitbed stratum of the Portland stone, the nature of the fossils is quite distinctive, see Figure 3.32.

All three main types of Portland stone (Roach, Whitbed and Base Bed) have their own characteristic texture and fossils. The stone is not pure white, but more of an 'off-white' colour and can sometimes be quite grey as a result of exposure to coal smoke and drainage water running over the surface. Readers unfamiliar with Portland stone will find that one quickly 'gets one's eye in' and the stone is easily recognisable from the fossil traces. There is a more detailed and excellent explanation of the fossils in the different types of Portland stone at the University College London website, www.ucl.ac.uk/~ucfbrxs/Homepage/walks/PortlandFossils.pdf.

3.4.4 Ham Hill stone

Next we illustrate Ham Hill stone, which is from the west of England, and is, like the Portland stone just described, a Jurassic limestone. It is readily identifiable by its rich honey-gold colour and clearly defined bedding planes which are accentuated by weathering. The overall appearance of a Ham Hill stone building is illustrated in Figure 3.33 by the front of the Lloyds Bank building in Salisbury, England.

Because of the quarry location, this stone is used mainly in the south-west of England and in nearby counties, but has also been used for buildings in Cambridge and Ireland. The attractive appearance of the Ham Hill stone can be seen in Figure 3.33 together with the clear evidence of bedding in the pillars and pilaster in Figure 3.34. The bedding structure is further shown in Figure 3.35 and the microstructure in Figure 3.36.

Leary (1983) notes that this stone is yellow–brown in colour with a mottled appearance due to the presence of iron and that many pieces of shell are evident. These characteristics

Figure 3.33 The façade of Lloyds Bank in Salisbury town centre, England.

are shown in the Figure 3.36 close-up photograph. Thus, Ham Hill stone is easily recognisable but has not been so widely used as the Portland stone. The more recent testing results from the UK Building Research Establishment indicate that the Ham Hill limestone has good weatherability and especially good frost resistance.

3.4.5 Kentish ragstone

Our third example of a building limestone, the Kentish ragstone, is a hard, blue-grey limestone from the east of England (the term 'ragstone' meaning a hard, coarse, rubbly or shelly stone). Leary (1983) describes this building stone as "a glauconitic sandy limestone from the Lower Greensands of Early Cretaceous age." The word 'glauconitic' means containing glauconite, a dull-green granular mineral of the mica group. As its name implies, Kentish ragstone is not a 'sophisticated' building stone but, because good building stone is not easily

Figure 3.34 Ham Hill stone pillars and a pilaster on the façade of Lloyds Bank, Salisbury.

Figure 3.35 Close up of the Ham Hill stone showing cross-bedding. Note that this stone has been placed upside-down relative to its original *in situ* bedding. This is evident from the relation between the cross-bedding and the two main horizontal bedding planes: the cross-bedding meets the upper main plane asymptotically and is truncated against the lower bedding plane. (See Section 3.5.3)

Figure 3.36 Close up of the Ham Hill stone microstructure showing small shell fragments.

available in Kent, nor indeed in the wider south-east of England, the ragstone has been used extensively in the area, especially for the Tower of London and for churches (see also Fig. 1.1 of the Jewel Tower in Westminster). It was used by the Romans, also in the Middle Ages, and more recently for buildings in London. We illustrate the use of Kentish ragstone for the building of Rochester castle, Figures 3.37–3.39.

The 12th century Norman tower keep at Rochester Castle was built by Gundulf, Bishop of Rochester, in 1078 and protected a crossing of the river Medway. The castle saw military action for the last time in 1381 when it was captured and ransacked during the Peasants' Revolt. The building of Westminster Abbey in the 1240s also required large quantities of ragstone—with the result that local supplies were commandeered for that purpose: a royal command decreed that "no Kentish ragstone shall be carted to London for any other purpose until the Abbey is built." Note that ragstone acquired its name from the quarrymen who so named it because it would break along ragged edges, Figures 3.38 and 3.39.

3.4.6 Ancaster stone

The fourth example limestone is Ancaster stone, an oolitic Jurassic limestone from Lincolnshire, this being our most northern illustrative limestone example. The term 'oolitic' means containing ooliths—which are small, spherical, calcium carbonate particles of the order of 0.25–2 mm diameter. The term derives from the Greek for 'egg' and the oolites form by the

Figure 3.37 The majestic Rochester Castle Keep built of ragstone, Kent, England.

Figure 3.38 Kentish ragstones, with a flintstone (see Section 3.6) at the centre of the photograph.

Figure 3.39 Staircase inside the ruined Rochester Castle.

precipitation of calcium carbonate around a small fragment of a shell or sand grain. Water currents can form cross-bedding in these oolites in exactly the same way as they do in sandstones (see Section 3.5.3) and it is the cross-bedding that gives the Ancaster stone its characteristic texture, e.g., Figures 3.41 and 3.42. Like many of the building stones in Britain, this limestone has been used for centuries by the Romans and Saxons through to the present day. The stone can be found in all the principal churches in Norwich, and has been used for the church and castle at Newark, and in buildings in Cambridge.

Our 'reference building' for this limestone is St. Albans Abbey in Hertfordshire, Figures 3.40 and 3.41, where Ancaster stone was used for the west front restoration by Lord Grimthorpe in the 19th century. Other examples are the castle, Figure 2.30, and St. Mary Magdalene church, Figure 3.42, both in Newark, Nottinghamshire. It can be seen from the relatively recently cleaned front of St. Albans Abbey, Figure 3.41, that the stone has a distinctive appearance because of its characteristic sedimentary texture.

The cross-bedding is accentuated by alternating layers of relatively light and dark layers, Figure 3.41. The light layers are probably fine muds settled out of suspension and draped

Figure 3.40 The Ancaster stone façade of St. Albans Abbey, Hertfordshire, England.

Figure 3.41 Ancaster stone on the cleaned façade of St. Albans Abbey, Hertfordshire, England.

Figure 3.42 Block of weathered Ancaster stone, exterior of the Church of St. Mary Magdalene, Newark-on-Trent, England.

over the cross beds during times of low current activity. Resurgence of current activity deposits more cross beds above the drapes of mud—and the process repeats.

The block of Ancaster stone from the exterior of the Newark church shown in Figure 3.42 illustrates the characteristic weathering nature of this building stone as influenced by the weaker character of the lighter beds, i.e., the selective weathering of the fine muds in preference to the more resistant oolitic limestone which makes up the majority of the rock structure.

* * * * *

In this Chapter 3 on recognising different types of building and decorative stones, we have so far described and illustrated granites and volcanic stones and now limestones. In this last limestone section, we have explained the characteristic natures of Portland stone, Ham Hill stone, Kentish ragstone and Ancaster stone. Recalling the earlier analogy referring to the ease and difficulty of recognising different species of flowers and birds, the limestones just described here are analogous to the daisies and robins of the natural world. There are, however, many variations on the limestone theme and the reader should not be disappointed if the stone in a particular building is not directly identifiable, or even whether it is self-evidently a limestone or a sandstone, because there is a wide spectrum of sedimentary stones containing different percentages of calcium carbonate and it may not be clear whether the

50% level which defines a limestone has been crossed. Hopefully, however, the four limestones that have been highlighted will now be readily recognised.

3.5 SANDSTONES

A sandstone is a sedimentary rock containing about 90% quartz, silicon dioxide, SiO_2, in the form of sand-sized particles.

3.5.1 Introduction

Unlike the limestones which we have just described (and which consist mainly of calcium carbonate, $CaCO_3$), sandstones consist mainly of quartz particles, SiO_2. These particles, originating from the weathering of igneous, sedimentary and metamorphic rocks, are transported by water and wind to areas where they accumulate as water lain sands or wind generated dunes. They there become cemented into a sandstone. Sometimes, the particulate nature is evident to the naked eye but the use of a hand lens will enhance one's appreciation of the microstructure. The particle sizes of the sandstones vary from the very fine to the very coarse, i.e., from 1/16 mm to 2 mm, and may be well sorted or poorly sorted in terms of grain size. If the particle shape is angular, the stone is termed a grit; otherwise it is termed a sandstone. The colour of the stone depends on the colour of the grains and the cementing mineral: white for a pure silica sandstone and brown/red when iron is present. Like the limestones, the sandstones within the UK also have a wide distribution through the geological ages from the Precambrian to the Cretaceous. Particularly in terms of their being used as building stones are the Old Red sandstone of Devonian Age and the New Red sandstone from the Permo-Trias.

The report on the building sandstones of the British Isles, which was prepared by E. Leary of the Building Research Establishment in 1986, provides a wealth of information on the different sandstones, identification features and associated reference buildings. Although it is generally easy to identify a specific building stone as being a sandstone because of its particulate structure and often the presence of 'cross-bedding' (see Section 3.5.3 and Fig. 3.56), it is not so easy to recognise specific sandstones, so we now illustrate four typical sandstone buildings, discuss how to establish 'which way up' a sandstone block is, and finally explain the mysterious feature of the yellow–brown coloured 'rings' often observed in sandstone building stones. The characteristic nature of a red sandstone building is illustrated in Figure 3.43 by the former Caledonian Hotel in Edinburgh, Scotland.

3.5.2 Examples of sandstone buildings

There are many sandstone buildings throughout Britain and they often have a characteristic red appearance as a result of their iron oxide (rust) content. Here we illustrate a few which are striking in their architectural appearance: the Edinburgh College of Art, the post-modern Number 1 Poultry building in London and the National Museum of Scotland in Edinburgh plus the Yorkstone widely used in urban pavements.

Edinburgh College of Art

Figure 3.44 shows the Edinburgh College of Art with its Scottish sandstone and, in Figure 3.45, there is an example of a pair of 'mirror matched' building stones from the front

Figure 3.43 Sandstone front of the old Caledonian Hotel in Edinburgh, Scotland.

Figure 3.44 Edinburgh College of Art, Old Red sandstone.

Figure 3.45 Two 'mirror matched' sandstone building stones, reminiscent of veneered furniture, from a stone block which has been cut in two, Edinburgh College of Art, Scotland.

Figure 3.46 Lettering in the sandstone façade of the Hunter Building, Edinburgh College of Art, Scotland.

of the Hunter building. The lettering at the front of this building, part of the Edinburgh College of Art, is shown in Figure 3.46. (We discuss stone lettering in Section 5.5.)

Number 1 Poultry, London

The words 'Number 1 Poultry' are an address in London: the street name is Poultry, originating from the poulterers' stalls which used to be there. The current building at this address was completed in 1997, is of post-modern design and incorporates sandstone cladding. The term 'post-modern' can be confusing and so we note that the Wikipedia entry for 'post-modern' architecture explains that,

> The characteristics of postmodernism allow its aim to be expressed in diverse ways. These characteristics include the use of sculptural forms, ornaments, anthropomorphism and materials which perform trompe l'oeil. These physical characteristics are combined with conceptual characteristics of meaning, including pluralism, double coding, flying buttresses and high ceilings, irony and paradox and contextualism.

Given that these concepts may not be too self-evident, e.g., 'irony', it is perhaps not surprising, therefore, that the Prince of Wales compared the 1 Poultry building (Figs. 3.47–3.50) to a 1930s wireless. Conversely, the building is said by some to be the most important

Figure 3.47 The 1 Poultry post-modern building in London, faced with bands of red and buff sandstone.

post-modern design in Britain. (We discuss the architectural aspects of stone buildings in Chapter 6 and explain post-modernism further in Section 6.11, which is titled "Post-modern architecture explained: the case of the bundled pilasters.")

Figures 3.48 and 3.49 provides a closer look at the sandstone facing stones. Siddall (2013) explains that the red sandstone is Wilderness Red, Old Red sandstone from the Forest of Dean, and that the buff coloured sandstone is from the Jurassic Helidon Sandstone Formation in Queensland, Australia, silicified by volcanic fluids from a basalt eruption. Sidall also explains that this eruption was responsible for the formation of fluids which caused the banding evident in Figures 3.48–3.50. The formation and appearance of such 'liesegang rings' in sandstone, which are not related to the bedding, are explained in detail later, in Section 7.3, where the majesty of Durham's Cathedral is described, together with the dramatic manifestations of liesegang rings in the Cathedral's weathered sandstone framework.

Edinburgh's National Museum of Scotland

The relatively new Museum of Scotland, Figure 3.51, located in Edinburgh's Chambers Street and which was opened in 1998, is faced with Moray sandstone, a Permian sandstone, from the Clashach Quarry. It is somewhat reminiscent of the 1 Poultry building just described, including the Prince Charles connection, in that he resigned as patron of the Museum in protest at the lack of consultation over its modern design. However, in the years since its construction, the Museum's design has become generally accepted, and in 2011 it physically merged with the adjacent and newly refurbished original 1861 Victorian museum building to become the National Museum of Scotland.

The decorative liesegang ring characteristics of the sandstone are clearly evident both outside and inside the Museum, Figure 3.52, and fit in with the post-modern architectural style.

Figure 3.48 The two types of sandstone in the exterior walls of the 1 Poultry building in London.

Figure 3.49 Sandstone cladding blocks on the 1 Poultry building. Note the granite blocks used at ground level because granite is more resistant to any flowing or standing rainwater, as was explained in Section 3.2. These blocks effectively act as a damp-proof course preventing water seeping up from the pavement and staining the more porous sandstone.

Figure 3.50 Enlarged image of the sandstone block from Figure 3.49 showing the 'liesegang ring' effect.

Figure 3.51 The National Museum of Scotland in Chambers Street, Edinburgh.

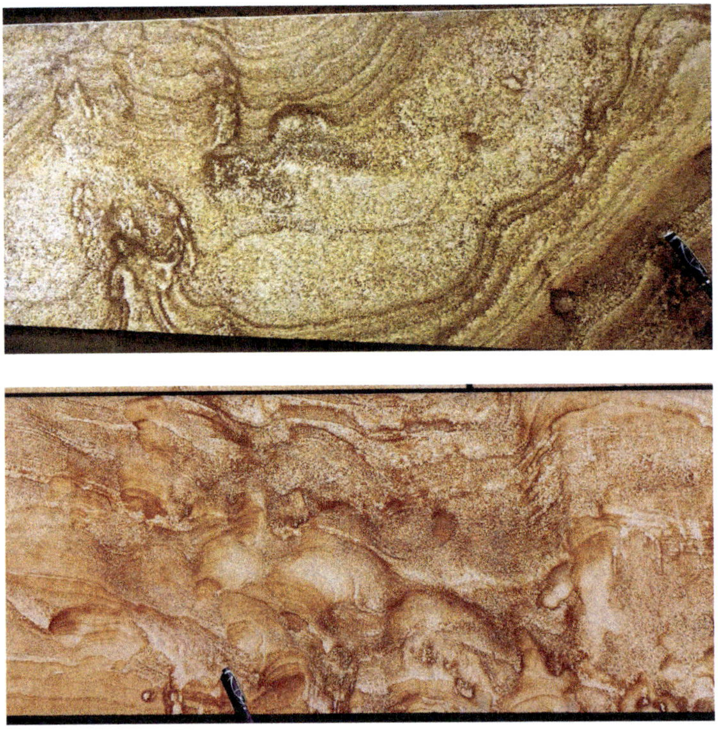

(a)

(b)

Figure 3.52 Two examples of liesegang ring formation in the building stones of the National Museum of Scotland, see Figure 3.51.

Yorkstone pavements

Yorkstone is a Carboniferous sandstone widely used throughout Britain as a paving stone and is easily recognised by its distinctive structure and colouration patterns, as indicated by the examples in Figures 3.53 and 3.54 from Salisbury and Liverpool.

Figure 3.53 Yorkstone pavement, Salisbury, England.

Figure 3.54 Pedestrian studs in wet Yorkstone pavement slab, Liverpool, England.

3.5.3 Which way up is the sandstone?

When studying sandstone blocks in buildings, it is satisfying to be able to establish which way up the building stone has been placed—is it the same way up as it was when the sandstone bed was originally deposited, or has it been turned upside-down? In this section, we explain both the original sandstone bedding process and hence the method by which the 'way up' can be established. The section concludes with a set of examples in a "Which Way Up?" quiz for readers.

During the original depositional process of sand particles falling on a sand dune or on the sea bed, the layers of particles are progressively built up. However, if this sequence of layers generated by the depositional process is interrupted by disturbances, such as alterations in the direction or intensity of wind or water flow, then the geometry of the layers is also interrupted. The resultant bedding structure then enables us to establish which way up the building stone has been emplaced. Is it the 'right way up' or is it 'upside-down'—relative to its geological origin? For example, looking at the lower central sandstone block in Figure 3.55, is this bedding the same way up as it was during the rock's formation? The answer is that the block is upside-down—which is established by the fact that the upper bedding in the block has been truncated (cut off) by the lower layers during deposition due to some disruption in the depositional process, see Figures 3.56–3.58.

As we have noted, sandstone is originally formed either on dry land (subaerially) or underwater (subaqueously). Due to winds at the surface and currents underwater, dunes are formed—and it is through the characteristic geometry of these dunes that we can establish the 'way up' for an *in situ* rock mass and for a specific sandstone building stone. The mason, usually unconcerned with the geological niceties established millions of years ago, may have inserted such a stone the right way up or upside-down in terms of its original geological orientation. The following text, explaining how the right way up can be established, is followed by a "Which Way Up?" quiz using photographs of four sandstone building stones.

Figure 3.55 Cross-bedding in sandstone visually enhanced by wind and rain erosion at Sanquahar Castle ruins on the Southern Upland Way (long distance walk), by Sanquahar town on the river Nith in Dumfries and Galloway, Scotland.

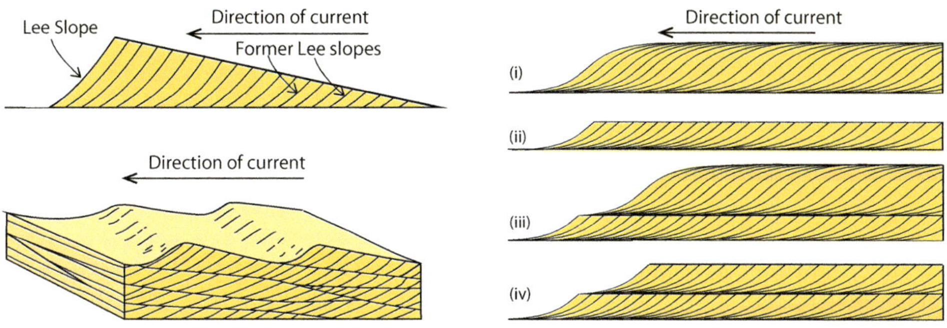

Figure 3.56 The sandstone depositional process and the formation of cross-bedding. Top left: profile through a dune showing the direction of the wind or water current generating it and the internal layering of the sand deposits. Lower left: block diagram showing the superposition of several dunes together with the regional bedding (gently inclined) and the cross-bedding (steeply inclined). Right: (i) the deposition of a cross-bedded sandstone layer, (ii) its erosion; (iii) the deposition of a second sandstone layer and (iv) its erosion.

Sands, whether accumulated on land or underwater, can become lithified to form sandstones. During their accumulation they develop a variety of features, some of which are common to the two types of accumulations and others which are different. Because these features are often displayed on sandstones, a brief outline of their formation is presented below.

In sand deserts, sand is transported by wind which forms dunes which move across the desert. A cross-section through a dune is shown schematically at the top left in Figure 3.56. The air current drives the sand particles from right to left, i.e., up the gentle slope (known as the stoss slope) to the dune crest when they tumble down the steep (i.e., lee) slope. Thus, there is erosion on the stoss slope and deposition on the lee slope. In this way, the dune migrates to the left and the bedding inside the dune is steeply inclined, i.e., parallel to the lee slope. Over time, dunes are blown or deposited on top of one another leading to the profile geometry shown at the lower left of Figure 3.56. It can be seen that within these dune accumulations there are two distinct orientations of bedding: the low angle bedding which marks the base of the dune, and the steeply dipping bedding parallel to the lee slope. These latter beds are termed 'cross-beds' and the whole geometry is known as cross-bedding.

Similar structures can form in sand bodies accumulated under water, as illustrated on the right-hand side of Figure 3.56. In the sketch marked (i), water carries sediments over the sediment–water interface to a small delta front where the sand is deposited. In these water lain cross-bedded bodies, both the stoss slope and the lee slope can be areas of deposition—with the result that the profile section of the cross-beds is an open 's' shape as shown. However, any subsequent pulse of fluid flow can lead to erosion of the upper surface as shown in (ii) resulting in the truncation of the cross-beds. This process can be repeated, leading to the typical cross-bedding profile shown in (iii) and (iv).

In both the wind and water generated cross-beds, the lee slope faces the direction in which the current of wind or water moved and the truncated surfaces of the beds are the upper surface. In addition, the concave (dished) face of the cross-beds is upwards. In this way, cross-bedding can be used to determine both the current and original 'way up' direction. Wind-formed (aeolian) dunes can often be distinguished from water-formed dunes (sub-aqueous) by their size; desert dune cross-bedding is generally much larger.

Inspection of the cross-bedding displayed by the lower, central sandstone block in the wall shown in Figure 3.55, shows that the upper cross-beds are truncated against the lower, more gently inclined beds. This indicates that the bed is upside-down compared to the orientation it had at the time of its deposition. Architects and masons and are not generally concerned with these details but, as discussed in Section 4.3, and illustrated in Figure 4.23, in order to take full advantage of the mechanical properties of a rock it is best to place it in the same orientation in a wall that it had during its formation.

In summary, the two 'way up' indicators are as follows.

- The first 'way up' indicator is the curvature of the bedding. If the bowl of the curvature faces upwards (concave upwards), the building stone is the right way up, i.e., it is the same way up as it was in the ground. Note also that the geometry of the cross-beds enables the direction of the wind or water current, the paleocurrent, to be established, see the right-hand side of Figure 3.56.
- The second 'way up' indicator is provided by the cut-off of the bedding surface, the truncation surface. The cross-beds truncate against the upper surface of a bed and merge asymptotically with the base of the bed, see Figure 3.56(ii).

Both these 'way up' indicators are illustrated in Figure 3.57.

Figure 3.57 'Way up' indicators for sandstone—both these indicators demonstrate that, in this example, the building stone is the right way up.

(a) (b) (c) (d)

Figure 3.58 A Which-Way-Up? sandstone quiz, see text. Can you establish which way up these four building stones have been placed? *The answers can be found at the end of this chapter.*

The set of four photographs in Figure 3.58 are presented as puzzles for readers to determine which way up these building stones have been placed, using the two indicators described above. Are the building stones the same way up as they were when they were formed, or have they been turned upside-down?

3.6 FLINT

Flint, crypto-crystalline silica, SiO_2, is one of the most imperishable materials on the Earth.
B.C.G. Shore, *Stones of Britain* (1957)

Sponges appear to have contributed most of the silica that now appears in the Chalk as flint. The relatively soluble opaline silica originally distributed through the Chalk has since been dissolved by percolating waters and redeposited in the insoluble form of flint.
Arthur Holmes, *Principles of Physical Geology* (1965)

Flint is one of the most intriguing of the building stones: it is chemically the same as quartz and glass; it occurs in irregular lumps; it is extremely hard; and it is highly resistant to deterioration. The soluble silica of which flint is formed originates through the chemical weathering of silicate materials. The silica is then transported by water and utilised for example by sponges. In life, the soft, porous, collagen sponges we are familiar with in our bathrooms are hard because they contain millions of brittle spicules (e.g., needle like rods), Figure 3.59(a), embedded in their bodies.

Many such spicules became fossilised in the Chalk strata of the Cretaceous period around 100 million years ago. Silica, being an amorphous, i.e., non-crystalline, form is more susceptible to solution than the crystalline quartz shown in Figure 3.59(b). The embedded spicules were dissolved in the groundwater and deposited in small solution cavities in the chalk as flint, itself a dense, cryptocrystalline variety of quartz, slightly translucent and almost opaque. Because the cavities were irregular in shape, so the resultant flint nodules have an irregular geometry, Figure 3.59(c). Moreover, the flints often have a fused white coating from the chalk cavity in which they were formed.

Figure 3.59 illustrates both (b) quartz crystals and (c) an irregular nodule of flint with an embedded Cretaceous fossil sea urchin—to emphasise that they are both a manifestation of silicon dioxide, SiO_2. Note that the fossil has retained its original shape through the gradual percolation of siliceous solutions through its original calcium carbonate shell, and the irregular shape of the nodule reflects the shape of the cavity in the chalk strata which was filled by the mobile silica. Thus, the irregular shapes of flint nodules are a direct indication of the irregular shapes of the cavities in the chalk strata in which the silica was deposited. Also, the ability of the silica to travel via groundwater explains why flint is often observed in coastal chalk exposures to be not only in horizontal veins, i.e., parallel to the bedding, but also in sub-vertical veins. This is because the groundwater was able to travel along all types of fractures in the chalk.

In ancient times, flint could not only be harvested from such coastal exposures but also in inland mines in the Middle and Upper Chalk strata, Figure 3.60, such as at the Prehistoric flint mines at Grime's Graves near Thetford in Norfolk. Flint does not occur in the Lower Chalk strata and, as mentioned earlier, luckily or perhaps by a flint avoidance strategy, the Channel Tunnel from England to France was excavated in the Lower Chalk using tunnel boring machines with cutters which would not have survived repeatedly hitting the extremely hard flint nodules. Note that where the percolation of silica rich fluids forms silica nodules in other formations, such as other limestones, calcareous shales and sandstones, the term *chert* is used to describe these nodules.

The map in Figure 3.60 indicates the locations where chalk occurs in England and hence the wide distribution of the use of flint. The nodules can be used as a building stone—either 'as is', e.g., Figure 3.61(a) and (b), or they can be 'knapped' (shaped) to have square or rectangular faces, Figure 3.61(c). The latter requires considerable skill because flint, like glass, naturally fractures with a conchoidal (curved) surface. However, it is difficult to find flint nodules large enough to form the corner stones, quoins, of a building, which is why many parish churches in the south-east of England have circular flint towers.

From the photographs in Figure 3.61, we can see how unaltered flint pebbles have been used directly for the construction of walls, Figure 3.61(a) and (b), and that the flint has been skillfully 'knapped' to produce the rectangular blocks in Figure 3.61(c). In fact, flint, as found on the ground surface or mined in chalk strata has been knapped and used by mankind since the earliest times because of the stone's strength and because it can be shaped relatively

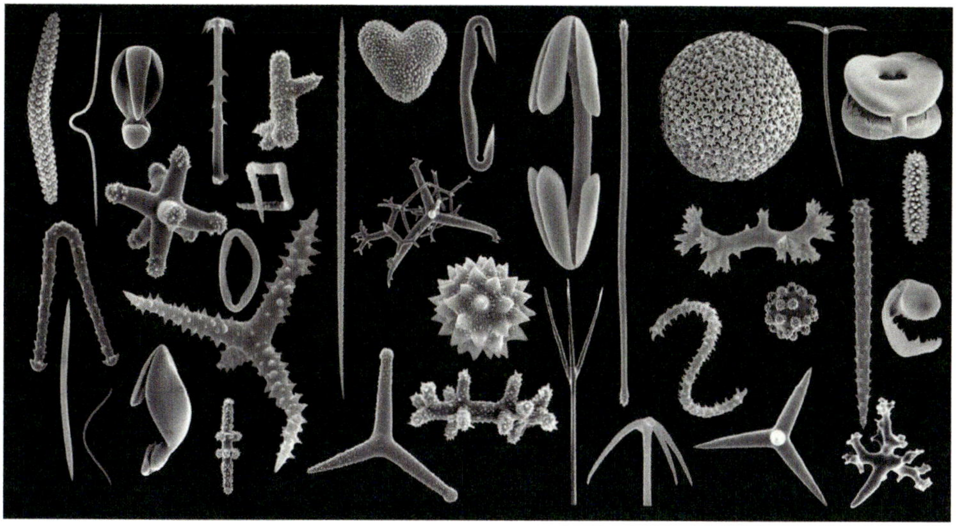

(a) Examples of microscopic sponge spicules.

(b) Crystalline quartz (SiO_2).

(c) Irregular flint nodule with an embedded sea urchin fossil (SiO_2).

Figure 3.59 All of the specimens shown above are made of silica, SiO_2. (a) Scanning electron microscope images of a selection of microscleres (minute sponge spicules) and megascleres (skeletal support elements in a sponge), not to scale, sizes varying between 0.01 and 1 mm. (b) Quartz crystals. (c) An irregular flint nodule (SiO_2) with an embedded sea urchin fossil attached (the latter is *Micraster coranguinum*, which means 'little star shaped like a heart'), from a field near Wheathampstead, Hertfordshire, England.

(a) From Rob W. M. Van Soest, Nicole Boury-Esnault, Jean Vacelet, Martin Dohrmann *et al.* (2012) Global Diversity of Sponges (Porifera). *PLoS ONE* 7(4): e35105. https://doi.org/10.1371/journal.pone.0035105 (Creative Commons Attribution License (CC by 2.5))

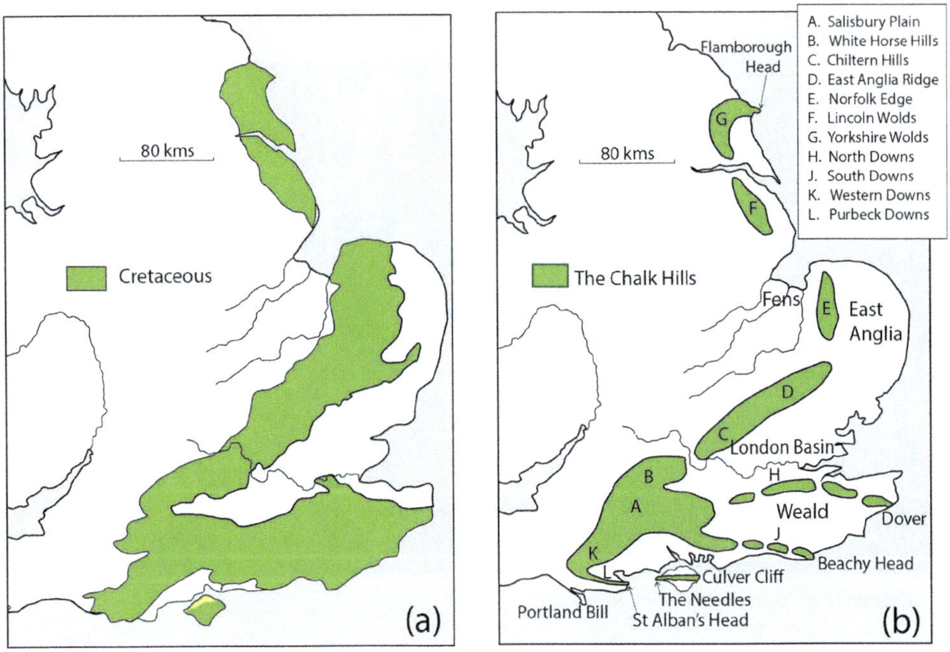

Figure 3.60 Left: The distribution of Cretaceous rock in England. The chalk, i.e., the strata containing the flints, is of Lower Cretaceous age. Right: The distribution of flint is controlled by the outcrop of the chalk which defines the various wolds, downs, hills, plains and edges.

(a) (b) (c)

Figure 3.61 Flint used as a building stone. (a) Flint nodules used directly 'as found'. (b) Use of flint collected from nearby fields to construct the Roman bath-house at Welwyn, Hertfordshire, England. (c) Flint nodules that have been 'squared' to a conventional building stone shape.

easily for a variety of purposes, e.g., before 10,000 BC for arrowheads and scrapers. Its use in other old buildings is shown in Figures 3.62 to 3.64.

Sir John Betjeman noted that "Rubble or uneven flints were not considered beautiful to look at until the nineteenth century. People were ashamed of them and wished to see their churches smooth on the outside and inside walls, and weatherproof." However, from the 19th century up to the present day, the use of flint as a decorative building stone has flourished; many church buildings have used a chequerboard (or diaper) style for the outside walls—as we illustrate in Figures 3.64 to 3.68.

One of the creative uses of flints is for gabions, a gabion being a wire cage filled with stones. These have proved useful for a variety of architectural and engineering purposes; for example, to create a temporary or a permanent wall—as shown in Figure 3.69 at the bus station in the town of Cromer in Norfolk, England and in Figure 3.70 at the Verulamium Park in St. Albans in Hertfordshire, England. As noted earlier, flint is highly resistant to deterioration so, if the wire cage is also resistant, then the structure can have a very long life.

Although we have mentioned that flint is highly resistant to deterioration, glass (having the same chemical formula) is not immune to the absorption of water in the long term.

Figure 3.62 Ruins of the old abbey in the Cathedral grounds at Bury St. Edmunds, Suffolk, England.

Figure 3.63 Flint wall in the grounds of Chichester Cathedral, West Sussex, England.

Figure 3.64 The porch of the 15th century St. Margaret's Church, Ditchling, East Sussex, England.

Figure 3.65(a) Chequer work using flint and Taynton limestone, Reading Minster of St. Mary the Virgin, St. Mary's Butts, Reading, Berkshire, England.

Figure 3.65(b) Close up of the chequer work using flint and Taynton limestone, St. Mary's Butts Church, Reading, Berkshire, England.

Figure 3.66(a) South entrance to St. Peters Church, Sheringham, Norfolk, England.

Figure 3.66(b) Close-up of the flintwork at St. Peters Church, Sheringham, Norfolk, England.

Figure 3.67 The entrance to the Methodist Church, Cromer, Norfolk, England.

Figure 3.68 Delightful conjunction of flint and terracotta within the entrance to the Methodist Church, Cromer, Norfolk, England.

Figure 3.69 Flint filled gabions forming part of the architecture of Cromer bus station, Norfolk, England.

Figure 3.70 Modern flint gabions in Verulamium Park near the remains of the Roman city of Verulamium, St. Albans, Hertfordshire, England. See also the dedication at the beginning of the book.

Figure 3.71 Old bottle exhibiting 'opalisation' of the glass.

Figure 3.71 illustrates how glass (SiO_2) changes its colouration when water of crystallisation is added ($SiO_2 \rightarrow SiO_2.nH_2O$), i.e., it opalises—in this case on the surface of an old glass bottle (Royal German Spa) retrieved from an early 20th century rubbish dump. The precious stone, opal, has the same chemical formula and the same type of colouration.

3.7 METAMORPHIC BUILDING STONES

Metamorphosis: Change of form by magic or by natural development.

Oxford English Dictionary

There are two fundamental types of metamorphic rock. The most common forms in response to a combination of heat and tectonic stress; such rocks are said to have been formed by '*regional metamorphism*'. The other type forms in response to an increase in heat alone and typically occurs adjacent to a hot intrusive igneous rock; consequently, such metamorphism is termed '*contact metamorphism*'. The main difference between the two types is that in contact metamorphism the newly formed mineral crystals display a random orientation, because there is no stress acting to cause any alignment. In addition, the crystals are locally developed around the intrusion where they form hard rims. Such rocks are termed hornfels. In contrast, and as the name implies, regionally metamorphosed rocks occur over large areas and often possess a marked mineral fabric.

Typical examples of metamorphic rocks are marbles, slates, schists and gneisses. Marble is produced by the metamorphism of limestone by heat and stress. The dominant mineral constituent of the limestone, calcium carbonate, recrystallises to form an interlocking mosaic of crystals, thus destroying any original porosity, primary sedimentary textures including bedding, and any fossil shell the rock contained. Thus, if a rock visibly contains shells, it should be termed a limestone and not a marble. However, non-geologists use the term much more loosely and many well-known limestones are termed 'marble' by architects and builders (e.g., the Purbeck marble, Fig. 3.27) even when the limestone is unmetamorphosed, as indicated by the fossil shell content.

Slate is formed by the metamorphism of mud, a fine-grained sediment, consisting predominantly of clay and quartz. The clay minerals recrystallise as fine-grained mica flakes which become aligned in response to the tectonic stress. This alignment imparts a planar fabric to the rock that enables it to split easily into thin sheets, as described in the next subsection. If the temperature at which the metamorphism occurs is higher, the mica flakes formed are larger and the resulting rock is termed a schist. Under extreme conditions of temperature and stress, the planar fabric becomes much coarser and the rock can develop a distinct banding; such a rock is termed a gneiss.

The two main metamorphic rocks used for building are slates and marble. These are the focus of the following section.

3.7.1 Slate

Slate is one of the most easily recognised building stones because of its ubiquitous use for roof and floor tiles across the whole of Britain. Its impermeability and long life result from its origin as a fine-grained metamorphic rock; and its usefulness as a rain resistant roofing material is because the stone can be split easily by hand into sheets. It might be assumed that the slate would split along the original bedding planes of the clay, but in fact it splits along the *cleavage* planes which have been formed during the metamorphism. This is generally not parallel to the bedding, see Figure 3.72(a). The slate extraction industry has been active for many years in Britain and large slate quarries have expanded over the centuries, especially in North Wales and Cornwall. The formation of the cleavage is shown in Figure 3.72(a) and an example of slate rock from a quarry is shown in Figure 3.72(b).

Figure 3.72(a) shows the relation between the folds and cleavage that form in a rock when it is deformed and metamorphosed in response to a tectonic compression. The cleavage forms parallel to the axial plane (i.e., the plane of symmetry) of the fold and is therefore termed an axial plane cleavage. Sometimes the mica flakes (the small rectangles in the figure) are aligned with their long axes parallel to each other within the cleavage plane, as indicated in the figure.

Following the explanation of Figure 3.72(a), the photograph of a small slate block, Figure 3.72(c), shows the key features: namely, the large cleavage plane with the through-going bedding (the set of horizontal lines) and on the joint surface indicated by the red dot, the

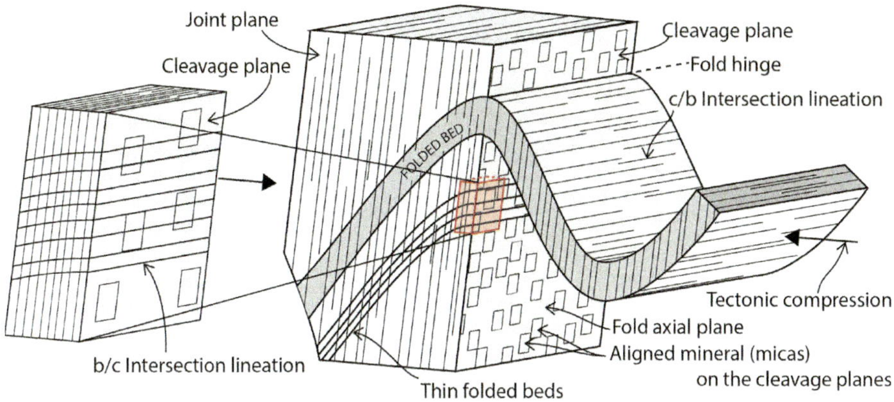

Figure 3.72(a) The process of slaty cleavage development. The c/b intersection lineation (right-hand diagram) is the lineation caused by the cleavage planes intersecting the bedding planes, and the b/c intersection lineation (left-hand diagram) is the lineation caused by the bedding planes intersecting the cleavage planes. This enlarged portion of the right-hand diagram shows the location within the fold from which the slate sample in Figure 3.72(c) originated.

Figure 3.72(b) Block of slate, *ca.* 200 mm across, illustrating the tendency of the slate to break into rectangular box type blocks. Both the top of the block and the left-hand side are pre-existing fracture surfaces, one a joint surface (the top) and the other a fault. Note the staining of the joint surface caused by iron rich groundwater, and the texture (lineation) of the fault caused by *in situ* shearing. The face of the block is one of the cleavage planes developed during metamorphosis.

Figure 3.72(c) Cleavage and bedding traces in a slate specimen. The larger face of the specimen is a cleavage plane, and numerous closely packed vertical traces of cleavage planes can be seen on the specimen face with the red dot. The horizontal parallel lines on both faces are the bedding planes. The coin is ~21 mm diameter.

Figure 3.73 Slate blocks used as building stones, Liverpool (section of wall ~ 0.5m across).

fainter vertical lineation of the cleavage planes. This sample was taken from the hinge region of a fold as indicated in Figure 3.72(a); we know this because it is the only location where the bedding and cleavage planes are at right angles to each other. Note that an example of slate used as a roofing material is illustrated in Chapter 8, Figures 8.5 and 8.6.

The faces of the blocks of slate in Figure 3.73 show the edges of the cleavage planes, which also form the top and bottom surfaces of the slate blocks. As noted above, slate splits easily along the cleavage planes into thick or thin sheets. In contrast, it is extremely difficult to split slate across the cleavage, i.e., at right angles to the cleavage. This is because the fractures have to cut across innumerable cleavage planes, each of which tends to inhibit fracture propagation. The result is that the cross-fractures are extremely irregular, as can be seen from the slate blocks used to build the wall shown in the figure. The top and bottom of the blocks are perfectly flat as they are controlled by the slaty cleavage, whereas the fractures at right angles to the cleavage which define the faces of the blocks are irregular.

3.7.2 Decorative marble—internal use

Marble: a metamorphic rock consisting of recrystallized calcite (calcium carbonate) and/or dolomite (calcium-magnesium carbonate).

In the previous sections of this chapter, we have described granites, volcanic stones, limestones, sandstones, flint and slate. Although, the emphasis of these earlier presentations has been on the building stones as structural building components, we have given examples where the stone is also decorative. In this section, we now concentrate on decorative marble *per se*.

Figures 3.74 and 3.75 show two examples of the use of decorative marble work, this use of stone being known as *pietre dure*, the Italian term for 'hard stones'.

The most extensive and spectacular use of marble in Britain is inside the Byzantine style Westminster Roman Catholic Cathedral, near Victoria Station in London, Figure 3.76. This Cathedral (not to be confused with the Anglican Westminster Abbey) stands on ground reclaimed from marshes by the old monks of the Abbey of St. Peter and was built from 1895–1903 with brick (12 million hand-made bricks) and stone. It has the highest (34 m) and widest (18 m) nave in the country and can accommodate 2000+ worshippers. Although located in a busy part of London, the Cathedral is set back from Victoria Street by a piazza

Figure 3.74 Old style rulers made with different types of marble held together by an internal wire rod.

Figure 3.75 Superb *pietre dure* marble inlay work from Florence in Italy (scale: ~0.7 m edge length).

Figure 3.76 Westminster Roman Catholic Cathedral, Victoria Street, London.

floored with Yorkstone and having attractive Rosso Verona limestone bollards. Decorative marble has been used throughout the interior of the Cathedral and can be especially studied in the several chapels on each side of the nave.

In fact, there are more than 100 types of marble sourced from 25 countries in the Cathedral. For those interested in visiting the Cathedral, we recommend the book by Rogers (2008) which not only describes the main cathedral marbles and their history, but also provides a photographic guide to the different marbles enabling the visitor to identify many of the specific marble types (Figs. 3.77 and 3.78).

The Cork Red, Lower Carboniferous pillars in Figure 3.77 are from Ireland and the marble is composed of pebbles set in a deeper red matrix, the red colour being caused by iron oxide.

Of particular interest in Figure 3.79 (in the top left position) is the marble crab—because it is embedded in a greenish marble which has a Scottish origin, being from a quarry on the island of Iona in the Inner Hebrides. Rogers (2008) notes that the Iona Green quarry is adjacent to the sea shore and that translucent green pebbles of the marble can be found on the beach, these being known as 'mermaid's tears' from the legend that a match between a mermaid and an Ionian monk was prevented by King Neptune and the Abbott.

Although we have been able to include here only a few examples of the marble used in Westminster Cathedral in London, hopefully these have illustrated the magnificence of the different types of decorative marble throughout the Cathedral. There are exciting histories

Figure 3.77 Cork Red marble pillars in the Norman arcading style colonnettes with brecciated marble forming the background wall in the Chapel of St. Patrick, Westminster Cathedral, London.

Figure 3.78 *Arni fantastico* marble from Tuscany intended to simulate the sea using two facing panels. Note that this is a similar technique to the wood veneer 'book match' system in which every other piece of veneer is turned over so that adjacent leaves are 'opened' like the pages of a book, creating a mirror-image pattern at the central joint line. See also Figure 3.45.

Figure 3.79 *Pietre dure* marble sea creatures in the floors of various chapels in Westminster Cathedral, London.

relating to the acquisition of these marbles: one story being that the architect had read about Verde Antico marble from an ancient quarry in Thessaly in Greece and he was not only able to find the specific quarry but, on re-opening it, was able to identify the furrows made by the wheels of the Roman carts in the 2nd century. After adventures transporting the stone from this quarry, it is now present in the Cathedral as eight columns of dark green marble supporting the gallery. The white capitals of some columns are Carrara marble from Michelangelo's quarry in Italy, a building stone which we discuss in Chapter 4 in the context of quarrying and in Chapter 8 in the context of 'deterioration' of cladding slabs when used externally.

3.8 BRECCIAS AND CONGLOMERATES

*A **breccia** (from the Italian for rubble) is a coarse-grained rock composed of **angular** fragments in a fine-grained matrix. A **conglomerate** (from the Latin for 'lumped together') is a coarse-grained rock composed of **rounded** fragments in a fine-grained matrix. This clear-cut distinction enables rapid identification of the two types of building stones.*

Breccias and conglomerates are not so widely used as the granites, limestones and sandstones previously described, but they are easily recognised and so have been included here.

3.8.1 Breccias

Our example breccia is the Welsh Radyr stone, a red Triassic sandstone breccia containing pebble-sized clasts of Carboniferous limestone. Because of its geological origin as a breccia (i.e., containing angular fragments), the stone has a rough texture and is difficult to cut; however, it is useful because of its attractive colourful appearance and its relatively good durability, one of the reasons for its use in the construction of Barry Docks. In Figure 3.80,

Figure 3.80 The Radyr stone breccia in the grounds of Llandaff Cathedral, Wales.

Figure 3.81 Red Radyr stone blocks in a Llandaff Cathedral wall, Wales.

we show the nature of the Radyr stone, with its contained angular fragments of limestone; a wall containing red Radyr building stones is included as Figure 3.81; and a severely weathered pillar of Radyr stone can be seen in Figure 3.82—all at Llandaff Cathedral (Eglwys Gadeiriol Llandaf) in Wales.

3.8.2 Conglomerates

A conglomerate is a coarse-grained sedimentary rock composed of large, sub-angular to rounded clasts (fragments) cemented in a matrix. The example in Figure 3.83 is of a Brno conglomerate (Czech Republic). The individual pebbles (the clasts) although slightly rounded by transportation in rivers, are still relatively angular. They are made of quartzite, a metamorphosed sandstone. In such rocks, the original sand (quartz) grains have been recrystallised and the original texture of the sandstone, which was made up of sand grains between which were pores, has been lost. The individual grains now form an interlocking mosaic of irregularly sized quartz crystals with very low porosity. Note that the matrix that surrounds the large-scale clasts is itself a smaller scale quartzite conglomerate.

The density of the larger clasts varies across the sample: where it is high, the clasts touch each other to form a 'clast-supported' conglomerate; where the density is lower, the clasts are not contiguous and the texture is described as 'matrix supported'. The matrix of the smaller scale quartzite is a sand with a quartz cement—so the rock is made up almost entirely of quartz. As a result, this conglomerate can be significantly resistant to weathering, like the millennia old Stonehenge monument trilithons in England (highlighted in the book's Frontispiece) which are a sandstone in which the pores have been infilled with a siliceous cement.

However, the Brno conglomerate is difficult to shape for building purposes because of its extreme hardness and pebbly composition, see Figures 3.84(a) and 3.84(b), and so it tends to be used in irregular lump form, Figure 3.85.

Figure 3.82 Centuries old outdoor weathered Radyr stone breccia in a Llandaff cathedral pillar, Wales.

Figure 3.83 Conglomerate used as a building stone in Brno, Czech Republic.

Figure 3.84(a) Ragged surface of a split Brno conglomerate building stone.

Figure 3.84(b) Smooth surface of a split Brno conglomerate stone.

Figure 3.85 Wall of irregularly shaped conglomerate building stones, Brno, Czech Republic.

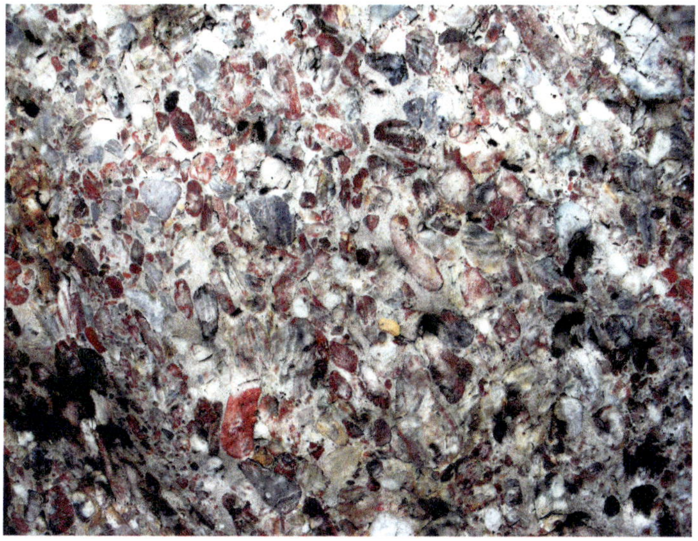

Figure 3.86 Conglomerate of rounded pebbles exhibiting a directional anisotropy, the sedimentary bedding, which runs from top left to bottom right. Specimen in the National Stone Museum, Wuhan, China.

The different example of a Chinese conglomerate shown in Figure 3.86 stimulates the following observations.

- The majority of the clasts making up this conglomerate are quartzite, despite the large variety of colours. Such conglomerates are termed monomictic.

- The clasts (pebbles) are sub-rounded to rounded indicating erosion as a result of being transported some distance from their source, most probably by water.
- The bedding is picked out by the stringer lens of smaller pebbles running from the top left to the bottom right of the photograph.
- There is a weak fabric in the rock parallel to the bedding defined by the alignment of some of the elongate pebbles.
- The conglomerate is clast supported, i.e., the clasts are in contact with each other.
- The matrix is a clean relatively fine-grained sandstone.

* * * * *

It should be noted that all the examples of conglomerates shown (Figs. 3.83–3.86) are quartzite conglomerates with a sand (i.e., quartz grains) matrix. The question naturally arises as to why so many conglomerates are of this composition. It will be recalled that sedimentary rocks are secondary rocks, i.e., rocks derived from other rocks by the process of erosion and transportation through a combination of water, wind and gravity. Thus, they are ultimately derived from primary, i.e., igneous, rocks, which are made up of a variety of minerals—conveniently divided into quartz, feldspars (potassium, sodium and calcium-aluminium silicates) and iron and magnesium silicates. These minerals are stable in the environment in which they formed in their parent magma, i.e., under conditions of high pressure and temperature; however, the majority of them are unstable in the low temperature, low pressure environment prevalent at the Earth's surface. But quartz is the exception. As these minerals are exposed to weathering and transported in rivers, they rapidly degrade and break down to the more stable clays that characterise many of the fine-grained sediments found in lakes, rivers and seas. This leaves behind the quartz grains and it is this mineral which survives river transportation and eventually ends up at the sea shore where it is deposited to form many of the beaches that characterise the edges of the world's oceans, the brown colour of sand coming from the iron oxide (rust) content.

In contrast, the breccias are much more varied in composition. Their angular clasts indicate clearly that they have not been transported far from their source and that the processes of weathering and erosion during their transportation will not have been of sufficient magnitude to destroy their less resistant clasts. As a result, many relatively unstable rocks can survive and be preserved in breccias; whereas, the long periods of transport that are associated with the clasts found in a conglomerate would result in their destruction.

* * * * *

A British example of a distinctive siliceous conglomerate is the 'Hertfordshire Pudding stone' shown in Figure 3.87, named after the English county north of London and the stone's resemblance to a currant pudding. When chalk strata were eroded, the resistant flint nodules contained became concentrated on the erosion surface where they were reworked by rivers. During their transportation they were rounded and sorted by size before being deposited and lithified to form a conglomerate with an iron-stained sand matrix.

Because the pebbles and the matrix are of the same material (silica, SiO_2), the Pudding stone fractures and is eroded equally through both, forming the surfaces shown in Figures 3.87. A section through this conglomerate is shown in Figure 3.88. However, if the silica matrix is less resistant to weathering, then the pebbles can protrude as illustrated in Figure 3.89.

Figure 3.87 Hertfordshire Pudding stone, Wheathampstead, Hertfordshire, England. The well-rounded geometry of the quartzite pebbles indicates that they have been transported a considerable distance by water.

Figure 3.88 Surface of a split sample of Hertfordshire Pudding stone, England.

Figure 3.89 Matrix erosion of Hertfordshire Pudding stone.

3.9 ARTIFICIAL STONES: TERRACOTTA, FAIENCE, BRICKS, CONCRETE, COADE STONE, GABIONS

For reasons of appearance or finance, an artificial stone may be chosen for a particular structure in place of natural stone. The most widely used such substitutes for decorative purposes are terracotta and faience, and for structural/substitutional purposes are brick, concrete and Coade stone.

- Terracotta: Porous, reddish, clay-based ceramic
- Faience: Glazed terracotta
- Brick: Kiln-fired sand and clay, typically 215 × 102.5 × 65 mm (UK size)
- Concrete: Hardened mixture of stone aggregate, sand, cement and water
- Coade stone: Type of terracotta in which several ingredients are fired twice

We discuss and provide examples of each of these in turn.

3.9.1 Terracotta

Terracotta is an unglazed, fired, clay material which has surprising durability and can remain unchanged for centuries. When deterioration does occur, it is caused by the original under-burning of the material, crystallisation of soluble salts or by frost action. One of the classic examples of terracotta work is the Chinese Terracotta Army at Xi'an, consisting of sculptures of thousands of soldiers and hundreds of horses created to protect the first Emperor of China after his death, and dating from the 3rd century BC, Figure 3.90. Given that these sculptures were buried and in existence for many hundreds of years, the terracotta itself has survived well, although many of the figures experienced mechanical damage as a result of being buried and then excavated.

Figure 3.90 Horses and soldiers in the First Emperor of China's Terracotta Army at Xi'an, China.

Figure 3.91 Terracotta building decoration in Crosshale Street, Liverpool.

Terracotta ornamentation of buildings has been very popular over the years. Two examples of its attractiveness are shown in Figures 3.91 and 3.92. The first of these, from a Liverpool building, demonstrates the appeal of the terracotta colour and texture; the second is a carved terracotta frieze on the façade of the Cutlers' Hall in London and shows cutlers working at their craft of knife-making and repairing.

On the larger scale, the Prudential Assurance Building, now known as Holborn Bars, in London is illustrated in Figure 3.93(a) and (b). This is a large, red, terracotta building

Figure 3.92 Finely carved terracotta frieze by the Sheffield sculptor Benjamin Creswick (1853–1946) on the façade of Cutlers' Hall in London.

in Victorian Gothic Revival style located in Holborn in London. It was designed by the architects Alfred Waterhouse and his son Paul, and was built during the period 1879–1901. The building was upgraded in the 1930s, especially the terracotta exterior so evident in Figure 3.93. Through the entrance shown in Figure 3.93(b) is Waterhouse Square, named after the architects. Note that Alfred Waterhouse was earlier the architect of the Natural History Museum in London, which we describe in the next section.

3.9.2 Faience: architectural glazed terracotta

The application of a glaze to terracotta forms a hardwearing, colour-fast surface with a gloss, eggshell or matt finish—such faience surfaces being popular cladding tiles in many London underground stations. With the variety of colours and textures that can be achieved, faience tiles indeed provide attractive building surfaces, see Figures 3.94 and 3.95. Also, the glassy nature of the surface not only reduces the development of grime but also makes the tiles easier to clean. Faience generally has a long life but, under adverse conditions such as thermal stress or shrinkage of the mortar used to bind the tile, a network of cracks can develop on the surface. Unlike sedimentary stone surfaces, glazed terracotta does not significantly age.

The portion of the 'tree of life' faience façade shown in Figure 3.95 is on the Bishopsgate Institute in London, which is a trade union centre of the General Federation of Trade Unions (GFTU)—a federation for specialist unions which includes the National Tile, Faience and Mosaic Fixers' Society.

The Natural History Museum on Cromwell Road in London is one of the best examples of architectural terracotta/faience and well worth a visit, see Figure 3.96. Following the Great Exhibition of 1851, several museums, including the Natural History Museum, Science Museum, Geological Museum and the Victoria and Albert Museum, were constructed in the Kensington area of London. The architect of the Natural History Museum was Alfred Waterhouse, a young architect from Liverpool. Construction work began in 1872 with the first phase of the Museum opening in 1881. The structure that Waterhouse designed has an internal iron frame and is clad externally and internally by faience tiles with natural history themed designs—the latter because he had been asked to harmonise the architectural decoration with the Museum exhibits. Living and extinct animals are incorporated into the

(a)

(b)

Figure 3.93 The Prudential Assurance Building in London. (a) View from the street. (b) A closer view of the entrance.

Figure 3.94 Semi-matte, blue-green faience tiles on the exterior of Holland House, London.

Figure 3.95 Portion of the light-coloured faience façade of the Bishopsgate Institute in London.

Figure 3.96 The Natural History Museum on Cromwell Road in London.

architectural details of the building, with extinct species in the east wing and living ones in the west wing. This Alfred Waterhouse building is designed in a mixture of Gothic Revival and 12th-century Romanesque style (see the architectural styles sections in Chapter 6). The powerful impact of the pillars at the entrance to the Museum can be seen in Figure 3.97 and two examples of the thematic tiles are shown in Figure 3.98.

Another faience example is the external surface shown in Figure 3.99 on the building of The Worshipful Company of Leathersellers—which is one of the ancient Livery Companies of the City of London and was founded by Royal Charter in 1444. The building is in St. Helen's Place, Bishopsgate, London. Above the ground floor, the façade is covered with a bespoke, mottled, glazed faience, reminiscent of leather in tone. One of the outstanding advantages of such a surface is the relative impermeability of its glazed surface which can be easily cleaned.

Figure 3.97 Spectacular decorative faience pillars at the entrance to the Natural History Museum, London.

Figure 3.98 Thematic tiles in the Natural History Museum, South Kensington, London.

Figure 3.99 Part of the new faience façade of Leathersellers Hall, London—the colour and texture being reminiscent of leather.

3.9.3 Brick

Historically and geographically, brick (typically containing sand, clay and lime and fired in a kiln) has been and is one of the most widely used structural materials for buildings. An example of the ancient use of bricks is shown in Figure 3.100 which shows part of the city wall surrounding the ancient city of Xi'an, China, home of the Terracotta Army, see Figure 3.90.

The brick wall in Figure 3.100(b) is interesting from two points of view. Firstly, the wall is built using the geometrical brick bond system now known as 'Flemish bond', see Figure 3.101. Secondly, these bricks are made from loess—which is a wind deposited sediment, is extremely fine grained and, importantly in the brick context, has a high clay content. Loess occurs all over the world, but particularly in Asia, and there is a loess plateau situated close to Xi'an. These loess bricks were fired in a kiln having been earlier dried in the sun and wind.

(a) (b)

Figure 3.100 The use of bricks in the buildings of the city wall of Xi'an city, the ancient capital of China.

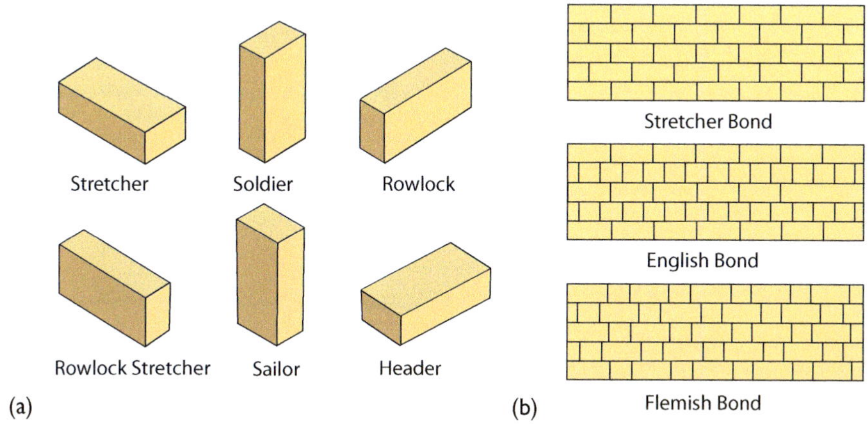

Figure 3.101 (a) The ways in which a brick can be laid when building a wall—the darkest sides being the face of the wall. (b) Three of the most common brick bonds.

Modern day bricks, made by firing clay, can be classified via their manufacturing technique and/or their appearance. **Facing** bricks are used where appearance is important, as in Figure 3.102; **engineering** bricks are required for structural purposes and where water resilience is important, e.g., railway structures and sewers; and **common** bricks used where neither appearance nor strength are important. The standard British brick became metricated in 1969 at a size of 215 × 102.5 × 65 mm, but if the nominal mortar width of 10 mm between the bricks is included then the working size is 225 × 112.5 × 75 mm. It is sensible to design the dimensions of a brick structure to be direct multiples of these dimensions—thus requiring a whole number of bricks.

Brickwork must be bonded so that the vertical joints between bricks do not lie directly above those in the course below, which would severely reduce the structural integrity of a wall, although it is recommended that the vertical joints should be above one another in *the next course but one*, which is the case for the Stretcher, English and Flemish bonds illustrated in Figure 3.101. Stretcher bond is the simplest and the most widely used, but English bond is stronger and Flemish bond is considered to be more decorative. The clays used for brick making in Britain have come from a variety of geological periods from the early Devonian (where the clay is derived from weathering of a hard shale) through to the current Holocene (alluvial, i.e., river deposited clays).

One of the best relatively recent examples of the architectural use of brick is the 'new' British Library which opened in 1998 near King's Cross Station in London. The building has a concrete frame and is clad inside and out in red brick—hand-made, sand-faced, dark Victorian Reds from Leicestershire which are laid in stretcher bond, as illustrated in Figure 3.101. The bricks at the British Library illustrated in Figure 3.102 are termed 'smiley' bricks because their surface texture is reminiscent of a smile. These bricks were chosen because they are made from the same clay as those used for the St. Pancras Station and Hotel adjacent to the east, Figure 3.103. Sir Gilbert Scott, the renowned 19th century architect who was responsible for the design of the Station (built during 1868–74), asked for Gripper's patent bricks from Nottingham to be used throughout that project, Figure 3.103. Glazed bricks, laid using an unusual bond, are shown in Figure 3.104 and an excellent example of the combined architectural use of brick and stone is illustrated in Figure 3.105.

Figure 3.102 Victorian Red 'smiley' bricks from Leicestershire laid in stretcher bond style as used in the British Library building, London.

(a)

(b)

Figure 3.103 (a) Brick and stone façade in the Gothic Revival St. Pancras International Station and Hotel to the east of the British Library. (b) Notice the roped window cleaners high up above the main entrance to the building.

Figure 3.104 An example of glazed bricks in London built with an unusual bond.

Figure 3.105 Example of the combined architectural use of brick and stone to excellent decorative effect at Bury St. Edmunds railway station in England.

Lettering in brickwork is more difficult than in stone because the distances between adjacent bricks tend to be smaller than in a stone wall and so the words often cross the boundaries. However, this has been successfully achieved in an interesting glazed brick wall bearing a quotation of Prince Albert's words describing his marriage to Queen Victoria on their 21st wedding anniversary in 1861: "How many a storm has swept over it and still it continues green and fresh, and throws out green shoots, from which I can acknowledge that much good will yet be engendered for this world"—as illustrated in Figure 3.106.

In terms of the deterioration of bricks, the type of surface flaking in Figure 3.107(a) and internal degradation in (b) is caused by weathering, specifically water penetration beneath the surface layer, compounded by freeze/thaw action through the seasons. However, the decay of bricks and brick walls over time can result, not only from degradation of the bricks themselves as in Figure 3.107, but also by structural collapse of the wall itself. The most important weathering processes directly affecting brickwork are frost action, crystallisation of soluble salts just below the brick surface, and expansion of the mortar when the brickwork remains wet for some time. The soluble salts are the sulphates of calcium, potassium, sodium and magnesium, the latter two being the most destructive. Note that, in the case of the weathering of sandstone, to be discussed later in Section 7.3, it is the hydrous calcium sulphate (gypsum) that contributes to the weathering of Durham Cathedral and castle. An additional and perhaps unexpected direction of attack on the mortar in brick walls can come from bees. These 'mason' bees like sunny walls and can make nests within cavities in the mortar between bricks. They do not eat the mortar but burrow into it.

In Figure 3.108, two examples are given where the problem is the structural integrity of the wall itself, rather than the bricks. On the left, is a through-going crack caused by differential building movement. A displacement transducer in the form of a sliding ruler is being used to establish whether the crack is growing. On the right is a crack traversing a modern brick wall, also caused by ground movement of the foundation. Note that, in both cases, such through-going cracks do not generally split the bricks: rather, they take the path of least resistance along the brick joints. Also, in both cases, there are subsidiary cracks around the main crack, indicating degradation of a zone, rather than just a single crack in the brickwork.

The brick/stone lined sewer tunnel shown in Figure 3.109 also indicates structural instability because of the longitudinal cracks in the roof. These occur because the weight of the

Figure 3.106 Lettering in Flemish bond brickwork, Howick Place, London.

(a)

(b)

Figure 3.107 (a) Brick spalling/flaking/exfoliation damage. (b) Internal brick degradation.

Figure 3.108 Structural degradation of brick walls through cracking. Left: Ancient wall at the city of Xi'an in China. Right: Relatively modern wall. Note that the cracks follow the path of least resistance, i.e., the mortar between the bricks—meaning that it is a structural failure, rather than a brick material failure.

Figure 3.109 Longitudinal cracking in the roof of a brick/stone lined sewer tunnel.

overburden above the tunnel causes a tensile (stretching) stress to be generated horizontally in the roof with its maximum value at the crown (top) of the tunnel.

3.9.4 Concrete

Although concrete is a manufactured material as distinct from a natural stone, we include mention of it here because concrete is used as a stone substitute and is a ubiquitous material in our continuously increasing metropolitan world. It plays a major part in the construction of buildings and other types of infrastructure such as roads and, in some cases, as a directly decorative material. The photographs in Figure 3.110(a) and (b) illustrate the use of concrete for the exterior of a building with a close-up of its pre-cast, polished concrete. Another close-up example of a decorative concrete texture is shown in Figure 3.111. Indeed, the textures in these figures are reminiscent of natural breccias which we described in Section 3.8.1 and which are made up of lithic (rock) fragments in a finer grained matrix. A different, but equally successful, exterior concrete surface texture is illustrated in Figure 3.112, highlighting just one of the many possibilities that concrete can offer.

Concrete has a long history of use, going back to the Romans, who constructed the Colosseum in Rome with concrete, brick and stone. Well-known buildings with concrete exteriors are the Guggenheim Museum in New York City and the Canary Wharf Tower in London.

(a) 

(b) 

Figure 3.110 (a) Concrete and glass exterior of 1, Coleman Street, London. (b) Close-up of the precast polished concrete.

Figure 3.111 Decorative concrete reminiscent of natural stone textures (see Fig. 3.86), Victoria Street, London.

(a)

Figure 3.112 Concrete exterior at 1 Moorgate Place, London.

(b)

Figure 3.112 (Continued)

Concrete is also effectively waterproof and used for such projects as the Grand Coulee Dam in the United States, where 20 million tonnes of concrete were used. Roadside pavements are also made of concrete but, for this application, a natural material such as Yorkstone is more traditional, as described in Section 3.5.2.

Concrete is manufactured by mixing cement, sand/gravel and water, the cement being made by heating a specific mixture of crushed limestone/chalk plus clay and sometimes water in a kiln. The wet concrete is poured into a mould and vibrated to remove entrapped air. The use of cementing materials to make artificial stone is very old: the ancient Egyptians used calcined (heated) impure gypsum, while the Greeks and Romans used calcined limestone but also learned to incorporate an aggregate. The most well-known cement in Britain is Portland cement, patented in 1824, which is made from a calcareous (containing calcium carbonate) stone such as limestone or chalk and from alumina (aluminium oxide) and quartz (silicon dioxide). The name 'Portland' comes from the fact that the cement resembles Portland limestone quarried in Dorset in southern England.

One of the major advantages of concrete over natural stone, apart from the reduced cost, is that reinforcement rods can be included in the mould so that the concrete's tensile strength is significantly enhanced. Natural stone does not have a high tensile strength (i.e., its resistance to being pulled apart), so its use is limited to beams with short spans because long stone beams are heavy, causing a tensile stress at the centre of the lower face of the beam and hence potential cracking and breaking of the beam. Thus, there is more structural potential when using reinforced concrete because the steel reinforcing rods can sustain the tensile stress—the success of which is evidenced by the many modern non-rectangular and large building spans.

Technical note on the compressive strength of natural stone and concrete

The uniaxial compressive strength of concrete (resistance to being crushed), being in the range 10 to 60 MPa, is similar to that of the weaker stone types. To have a feeling for a megapascal (MPa) one can start with Imperial units. Assume that a cube of stone which is one foot along each of its sides weighs 144 pounds. Its base then has an area of 12 × 12 inches so the stress the cube exerts on the base is (weight/area) 144 pounds/144 inches = 1 pound per square inch = 1 psi. The conversion of units from psi to MPa is 1 MPa ≅ 145 psi, so 1 MPa is the force/weight exerted at the bottom of a stone pillar 145 feet high, i.e., 44 m high. This also indicates that the vertical stress component in a natural rock mass increases in the order of 1 MPa for every 44 m of depth. The compressive strength of all stones varies from about 20 to 240 MPa, with the majority being in the 20 to 70 MPa range.

Needless to say, the strength of the concrete will depend on how well it is made, and the strength of a particular type of building stone will depend on its geological origin and subsequent environment but, in general and considering normal concrete, on the average stone will have a higher compressive strength than concrete. The strengths of building stones is discussed in Chapter 9 of our companion book *Structural Geology and Rock Engineering* (Cosgrove and Hudson, 2016).

3.9.5 Coade stone

The artificial building stone Coade stone is named after Mrs. Eleanor Coade (who was born in 1733 in England) and was used mainly for sculptures and door surrounds. The main advantage of Coade stone is its extreme resistance to rain penetration and frost damage—due to its manufacturing process which involved the ingredients of pipe-clay, flint, sand, glass and stoneware (ceramic ware that is fired in high heat, making it vitrified and non-porous), and a double firing process (van Lemmen, 2006). In fact Coade stone was sometimes referred to as Lythodipyra from the Greek *litho* (stone), *di* (twice) and *pyra* (fire).

The best known example of Coade stone is the lion sculpture mounted on Westminster Bridge in London (Fig. 3.113). Because of its relatively low cost (compared to Portland stone) and its excellent resistance to the weather, many building and garden ornaments, statues and tombs were made in Coade stone in the 18th and 19th centuries, e.g., the doorway surround in Figure 3.114. Sir John Soane (architect of the Bank of England in London) had a long association with the Coade factory and used the material for many of his works. The booklet by van Lemmen (2006) lists locations in the British Isles where examples of Coade stone can be seen.

3.9.6 Gabions

To finish this chapter on artificial stones, we should also remember ‘gabions’, which were mentioned earlier at the end of Section 3.6 on flint. A gabion is a wire mesh box filled with natural or artificial stones, the word gabion coming from the Italian *gabbia* for a cage or basket. The content of a gabion can be any stone or stone-like material that has a good resistance

Figure 3.113 The Coade stone lion cast sculpture on Westminster Bridge, London.

Figure 3.114 Doorway decorated with the architectural ceramic Coade stone at 28 Bedford Square, London.

Figure 3.115 A retaining wall made from gabions infilled with quartzite rocks.

to weathering, and the wire mesh frame should be galvanised wire or stainless steel wire. So, although the gabion is generally created using natural stone, as a structural building element it is an artificial building block. Examples of a wall with gabion components are shown in the dedication at the beginning of the book and in Figure 3.115.

Gabions have many advantages compared to dimensioned stone blocks. They can be filled with irregular stone pieces and hence are cheap, and they can easily be made to any suitable size. Moreover, when filled with impervious stones, such as flints, and contained within rust resistant wire, they have long-lasting properties and can be used underwater. Additionally, their construction, transport and emplacement usually involves less constructional disturbance than conventional building operations. Also, gabion structures can easily be taken down and reused in a different overall geometry. For decorative garden use, they can be filled or capped with stones matching adjacent structures. Gabions are not, however, to everyone's architectural taste.

* * * * *

Answers to the 'Which Way Up?' quiz, Figure 3.58

(a) Right way up. (b) Wrong way up. (c) Right way up. (d) Wrong way up.

Chapter 4

The life of a building stone

Quarrying and emplacement through to deterioration

4.1 INTRODUCTION

In Chapter 2 we described the geological background to the origin and variety of building stones. This was followed in Chapter 3 and in the geological context by descriptions and illustrations of the main types of building stones so that they can be more easily recognised. Now, we describe the life of a building stone—from initial quarrying, through preparation for a particular building, subsequent construction, and then the stone's long-term deterioration 'back to the Earth'. Further explanations of building stone deterioration and the inevitability of stone building ruins are presented in Chapter 8.

4.2 QUARRIES AND THEIR DISTRIBUTION IN BRITAIN

The locations of all quarries will naturally follow (a) the localised need for the building stones and (b) the positions of suitable geological strata, as illustrated in Figure 4.1, although nowadays it is much easier to transport building stones from more distant locations.

The active quarries in the British Isles are listed in the *Natural Stone Directory*, which is compiled and published annually via the *Natural Stone Specialist* magazine and is available from QMJ Publishing Ltd. The directory contains a wealth of information on building stones and quarries, including colour images of cut and polished samples of different granites, limestones, sandstones and slate. This is followed by a county-by-county listing of the active quarries, with details for each quarry relating to the rock type, colour, texture, technical information, buildings where the finished stone can be viewed, etc. Also included in the directory are listings of quarry operators, wholesalers, equipment suppliers, stonemasons and related organisations.

Although the content of the past and current Natural Stone Directories indicates a large and active building stone industry, there is evidence that the number of active quarries in Britain is declining over the years—although not necessarily the total output of building stone. In the Figure 4.2 graph, there is a comparison of the number of limestone, sandstone, slate and granite quarries in the years 1991 and 2017, i.e., over a 26-year period. For example, the number of limestone quarries has reduced from 312 to 120 in that period. No doubt, a variety of factors contribute to this overall reduction, including lower cost/tonne in the larger quarries and increased use of stone from overseas.

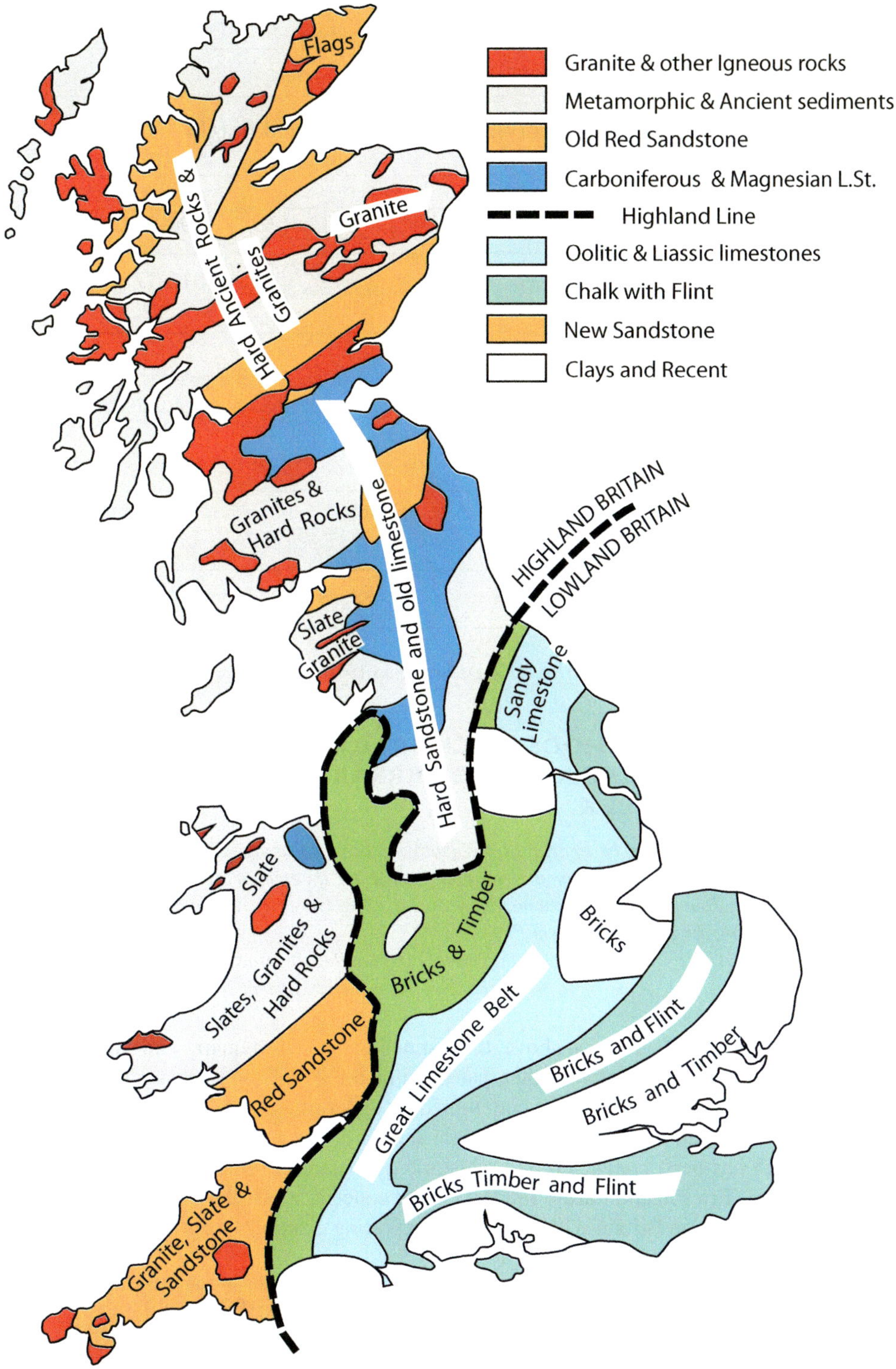

Figure 4.1 Map of localised British building stone. (Modified from Penoyre, J. and Ryan, M. (1975) *The Observer's Book of Architecture*. London: F. Warne and Co.)

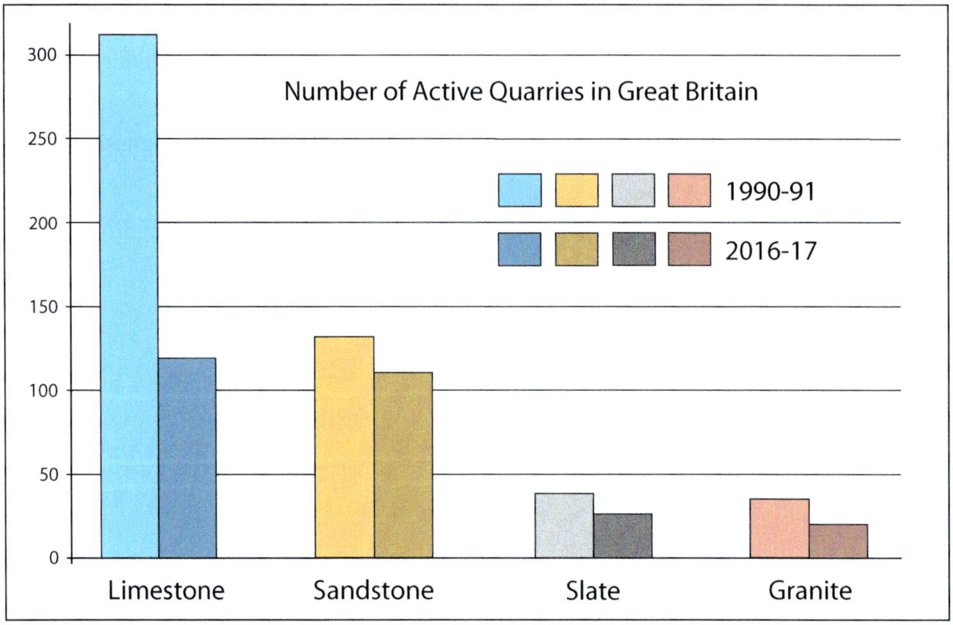

Figure 4.2 Comparison of active quarries in Great Britain in 1990–1 and 2016–17.

4.3 QUARRYING METHODS, ROCK FRACTURES, PORTLAND STONE AND THE CARRARA MARBLE QUARRY COMPLEX

A Mine is defined to be a certain Hole, Hollow place, or Passage digged in the Earth, from whence Metals or Minerals are by labour raised; for if common Stones only are found (as Marble, Touchstone,1 Freestone,[2] *etc.) we call them Quarries, and not Mines. And where Clays are digged we call them Pits.*

Sir John Pettus, *Fodinae Regales (the History, Laws and Places of the Chief Mines and Mineral Works in England)* (1670)

As indicated in the 1670 quotation above, the distinction between a quarry and a mine is that the term quarry is used when stone rather than metallic ore is being extracted, although there can be confusion when stone is also being mined by underground methods, as has been the case for Bath stone and Portland stone in Britain and Cararra marble in Italy. Although interesting research has been conducted on prehistoric quarries and lithic (pertaining to stone) production (e.g., Ericson and Purdy, 1984), here we concentrate on the current quarries and quarrying techniques in Britain with additional supporting examples from overseas.

4.3.1 The influence of the natural rock structure on quarrying methods

In Chapter 3, we used the geological basis of rock formation to present the different types of building stone, but it should be noted that over geological time rock masses have been subjected to a variety of natural factors during and after their formation, especially natural stresses in the rock which have changed their properties and which now have a significant influence on the quarrying process. In fact, there are three such geomechanical factors that play a role in determining the method of stone extraction and hence on the geometry of a quarry:

- the microstructure of the rock,
- the types of natural fracturing the rock mass contains, and
- the macrostructure of the rock, i.e., the overall geological formation.

At first thought, it might seem that the microstructure of the rock would not be an important factor, but consider the images in Figures 4.3(a) and 4.3(b) of a granite quarry. Notice that in both these figures there are some right-angled corners in the quarry geometry. This is not just because such right angles are convenient for quarrying, but because the granite has a microstructure with strength values which are different in the three mutually perpendicular directions—these being known as the rift, grain and hardway planes. The quarrymen take advantage of these microstructural planes to optimise the quarrying technique, with the consequence that the granite microstructure directly affects the overall quarry geometry, i.e., the quarry macrostructure. The same effect can be seen in Figure 4.4 in a different quarry where the 90° corners are also clearly delineated.

Figure 4.3(a) Part of a granite quarry. The linear and right-angled geometrical features are a direct reflection of the internal microstructure of the granite—which has different strengths in different directions, as illustrated by Case C in Figure 4.5. (Photograph by the authors and courtesy of Imperial College Press, London)

Figure 4.3(b) Close-up of the top left part of the quarry in Figure 4.3(a). Note the line of the tops of the metal wedges in the left foreground; these are used to split the granite. (Photograph by the authors and courtesy of Imperial College Press, London)

Figure 4.4 Marble quarry in Portugal illustrating the influence of the rock structure on the right-angled geometry of the quarrying technique.

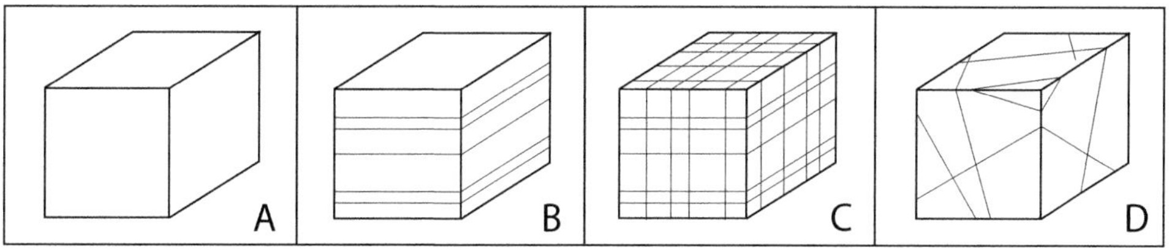

Figure 4.5 Four main types of natural rock fracturing.

The second bullet point at the beginning of this sub-section is the types of natural fracturing the rock mass contains. This is an important factor because the quarrying is improved if it operates in harmony with the natural fracturing. In Figure 4.5, four idealised examples of natural rock fracturing are shown. The rock mass denoted by A contains no fractures; the B example has a series of parallel fractures known as a fracture set; the C type has three fracture sets which are mutually perpendicular and which together form a fracture group; and the D type contains 'random' fractures.

The technical terms for the simplified geometries of rock masses shown in Figure 4.5 and caused by natural rock fracturing are as follows.

A **Isotropic**. No fractures, and the mechanical properties are the same in all directions, e.g., an unfractured granite rock mass or a massive limestone or sandstone.
B **Transversely isotropic**. One set of fractures, with the properties being the same in the horizontal directions, but different in the vertical direction, e.g., unfractured but bedded strata.
C **Orthotropic**. Three sets of mutually perpendicular fractures, the properties being different in the three perpendicular directions, e.g., a sedimentary bedded rock with two sets of fractures perpendicular to each other and to the bedding (generated by burial and subsequent exhumation).
D **'Random'**. Many sets of fractures in different directions so the properties are different in all directions, e.g., a rock mass which has been subjected to a succession of deformation events over geological time, each of which has induced a new set of differently orientated fractures.

Rock masses occur in all of these fracturing modes, including superposition of two or more modes. The Type A truly isotropic rocks, either considered on the large or small scales, are rare because of the geological mode of rock formation and subsequent fracturing which will generally induce some form of structural anisotropy. However, there are rock masses which have few natural fractures, such as the marble rock mass in the underground quarry shown in Figure 4.6.

A Type B rock mass, schematically outlined in Figure 4.5 with essentially one set of parallel or sub-parallel fractures, occurs in sedimentary rocks containing bedding planes which may cut a uniform rock type or mark the boundary between two distinct rock types. A Type C rock mass is a further development where *in situ* rock stresses have caused fractures mutually perpendicular to the strata, see Figures 4.7(a), (b) and (c). In these examples, the primary 'fracture' set, the bedding, combines with two secondary, i.e., stress-induced fracture sets,

Figure 4.6 Underground marble quarry. This type of quarrying procedure involves cutting out large blocks with saws, diamond encrusted wire or hydraulic 'flatjacks', leaving a series of flat marble faces. Note that lines in the marble have been caused by the quarrying procedure and are not fractures. However, occasionally a natural facture is encountered as indicated by the white arrow. Note the staining on this fracture face which has been caused by the flow of natural groundwater over geological time.

to form a regular fracture network which may have an important impact on the extraction of blocks of building stone from a quarry.

There are two factors that favour the formation of the regular organisation of the fracture sets that characterise Type C fracturing. The first factor is the existence of the primary bedding; this tends to influence the orientation of the natural stresses that develop in the rock so that they act either parallel to or at right angles to the bedding with the result that joints that form in response to such stresses will be either at right angles or parallel to the bedding. The second factor is the stress regime to which sedimentary rocks have been subjected during their history which has involved the addition and removal of the overburden stress.

In Type D fracturing (Fig. 4.5), the fracture sets appear more randomly arranged. An example from the Lake District in England is shown in Figure 4.8. Such patterns are likely to develop in the more massive rocks, such as igneous rocks, which have been subjected to several tectonic stress episodes and where the stress orientation is unconstrained by bedding. However, careful study of Figure 4.8 reveals that the fracturing is not as chaotic as

Figure 4.7(a) Type C fracturing in Carboniferous limestone, South Wales. The three fracture sets are mutually perpendicular.

Figure 4.7(b) Type C fracturing in a disused Niagara dolomite quarry at Lannon, Wisconsin, USA. Note the well-developed, horizontal bedding and the vegetation growing in the major vertical rock joints which outline this block.

Figure 4.7(c) Type C fracturing in the St. Bees sandstone, northern England.

Figure 4.8 Natural rock fracturing (Type D, *cf.* Fig. 4.5) resulting from successive phases of natural fracturing events (Lake District, England). Note, however, that elements of Type C (mutually perpendicular fractures) can also be seen in this photograph.

may appear at first sight because in fact there is a basic Type C geometry (note the presence of right-angled blocks in the figure) overlain by a variety of other angled fractures. Study of the fractures in the disused Cheesewring granite quarry face in Figure 4.9 also indicates that the geometry of natural rock fractures follow a pattern rather than being random *per se*. This subject concerning the origin and geometry of natural rock fractures is dealt with in detail in our more technical companion book relating structural geology and rock engineering (Cosgrove and Hudson, 2016).

In the rock mass examples in Figures 4.7–4.9, any bedding has combined with stress-induced fracture sets to form a fracture network that may have either a positive or negative impact on the extraction of blocks from a quarry.

There are more complex naturally occurring building stone geometries. One example is flint (see Section 3.6) which occurs as nodules in chalk formations and has been used for

Figure 4.9 Portion of the disused granite Cheesewring Quarry, Bodmin Moor, Cornwall, England.

thousands of years for making items such as arrowheads and skin scrapers, and as a building stone. It forms after the Chalk has been lithified, i.e., buried and compacted into a rock, and consists of non-crystalline quartz which is precipitated along bedding planes and fractures from silica rich fluids moving through the Chalk. The most well-known flint mining location is the Neolithic site of Grime's Graves in Norfolk, England, where there is a complex of 700–800 shafts providing access to tunnels and underground chambers in the flint rich Chalk stratum.

An example of a geological process that can lead to the formation of highly irregular blocks in a rock which do not split easily is the formation of lava flows. When lava flows evenly over a flat surface and cools, it can form regular blocks of volcanic rock whose geometry is defined by the horizontal lava base, the top, and the vertical cooling fractures, as occurs for example in the famous basaltic flows of the Giant's Causeway in Northern Ireland. However, when the lava flows over an irregular topography, the top and bottom surfaces of the flow will have different geometries and the orientation of the cooling fractures will vary. The resulting blocks of rock will have complex geometries which in terms of a suitable size or shape are awkward for use as a building stone. However these may be the only available, as is the case for Jeju Island in South Korea, see Figure 4.10.

4.3.2 Portland stone

> *'That', said the quarry foreman, 'is where St. Paul's came from!' I looked down toward the sea from a high cliff on the eastern side of the Isle of Portland. I saw a valley gouged out of the hillside: a dead, desolate wilderness; a cutting away of high stone cliffs as if some race of giants had scooped out the stone to its bed. . . . I realize now that no one understands London until they have explored the significant chasms [of the Portland stone quarry]. . . . Somerset House, St. Paul's, the Bank, the Royal Exchange, the Mansion House, the Law Courts, the British Museum, all the Wren churches . . . have left their caves, gullies, and gaps in the Isle of Portland. . . . I thought, not only of the buildings which Portland has already given to London, but of the London to be that slumbers still in darkness in the womb of this pregnant Isle.*
>
> From a description of a visit to the limestone quarry at the Isle of Portland on the south coast of England by H.V. Morton, in his book *In Search of England*, Methuen (1927), Penguin Books (1960)

The most famous British quarry supplies Portland stone from the Isle of Portland off the coast of Dorset in England. As indicated by the H. V. Morton quotation above, this stone has been used for centuries, especially in London and particularly for St. Paul's Cathedral, which we highlight in Section 6.12 when discussing the architect Sir Christopher Wren. This stone has particularly good strength and weathering properties, is relatively easy to extract from surface and shallow rock strata, is very suitable for decorative and lettering purposes, and has been used for many prestige projects. In addition to the 'plain white' variety of this stone, there are also visibly fossilised varieties, Figure 4.11. Since the days of the H. V. Morton quotation above, the easily available surface Portland stone has been extensively excavated and much of the production is now via underground mining.

The particular strata from which the different types of Portland stone are extracted occur in extensive horizontal seams like coal and the mining technique must ensure that the stone is extracted safely without prejudice to either the workers or indeed the mine itself. In coal mining, there are two types of extraction techniques: room and pillar mining, in which only

(a)

(b)

Figure 4.10 (a) An irregular lava flow devoid of any systematic fractures and (b) its use as a building stone, Jeju Island, South Korea.

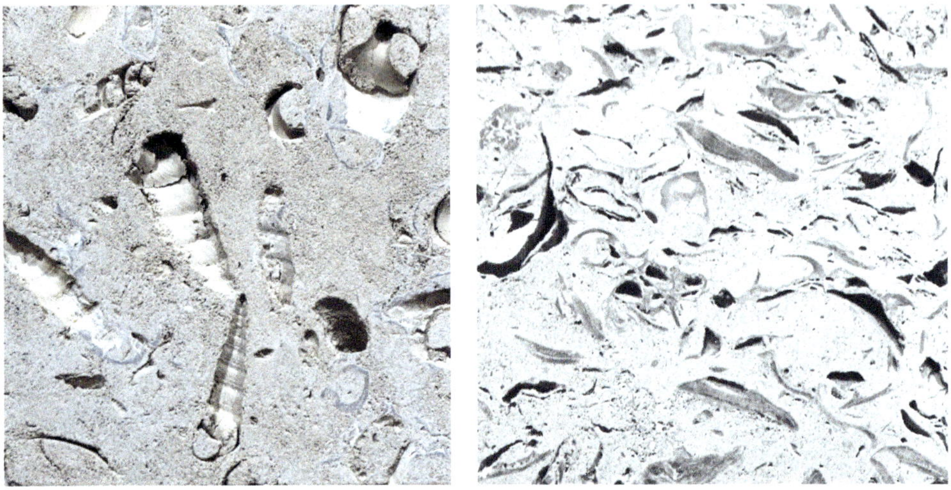

(a) Portland Bowers Roach. (b) Portland Grove Whitbed.

Figure 4.11 Two specimen samples of fossiliferous Portland stone (each ~80 mm width).

part of the coal stratum is removed leaving pillars to support the roof, and longwall mining, in which essentially all the coal seam is extracted by mechanical shearers and the roof collapses safely behind the working face as part of the mining process. However, the longwall mining method is not a practical option for mining building stone and so the room and pillar method is used. As the name implies, the stone is removed, usually on a square/rectangular grid geometry, leaving open rooms and pillars to support the roof. If an attempt were made to extract all the stone, the rock overburden would collapse and so the mine would have to be fully supported by artificial means, which would be expensive and impractical; however, a large percentage of the stone can be removed safely (up to 75%), especially as the strata are relatively shallow and so the vertical stress component in the rock is low.

To extract one of the blocks of stone, a chain saw type cutter (2 m long and diamond tipped) is used to cut slots into the limestone seam around the block to be extracted. Then, a flatjack, a thin steel envelope (known as a 'hydro-bag') is inserted into one of the slots near to an open face and pressurised with water. The large force thus induced causes a block to be separated from the seam, usually with little damage. By repeating this process, the mine advances, leaving the overall mine geometry as visible in Figure 4.12. To provide a safe working environment, rockbolts 2.4 m long are inserted vertically into the roof at regular intervals, Figure 4.13, in case there are any rock fractures within the rock mass that could cause rock blocks to fall out, or indeed for the strata above to separate and form a major hazard. The rockbolts reinforce the strata by mechanically connecting the immediate roof stone with the strata above. This mechanical reinforcement can also be used in a horizontal direction in the mine walls to strengthen regions where natural or artificially induced vertical fractures have developed, Figure 4.14.

This being an *in situ* rock mass, it will contain natural fractures as we have described earlier in this section; one such large natural fracture is highlighted in Figure 4.14 and, as indicated in the figure caption, it is a large, through-going fracture. The main purpose of the

Figure 4.12 View of the Portland stone underground mine where the limestone has been extracted using the 'room and pillar' method. The stone has been extracted leaving the pillars which are necessary to ensure the stability of the roof. (Courtesy of M. Godden, Albion Stone)

Figure 4.13 Inserting rockbolts in the roof to lock the roof stone to the strata above. Note the square arrangement of rockbolt heads in the roof. (Courtesy of M. Godden, Albion Stone)

Figure 4.14 A major vertical rock fracture intersecting the mine workings. Note the extent of the fracture—from the right-hand wall, across the roof, through to the central pillar, through to the wall in the distance on the left-hand side of the photograph—and the metal reinforcement secured by horizontal rockbolts which inhibits further fracture development. (Courtesy of M. Godden, Albion Stone)

metal reinforcement straps around such large and dilated fractures is to prevent any loose material lodged within the fracture creating a hazard by sliding into the workings. Although such vertical fractures can interfere with the quarrying/mining process, they do not generally cause major instabilities; however, the intersecting of gently inclined fractures within the mine would be of greater concern as these could cause blocks of rock to fall or slide into the excavated rooms. Also, the direction of the roadways and hence the orientation of the mine have been chosen bearing in mind the orientation of such major pre-existing fractures; moreover, the mine design is dynamic and such large fractures are contained within oversized pillars in order to maximise the mine stability.

The authors are grateful to Mark Godden of Albion Stone for his assistance in the compilation of Section 4.3.2.

4.3.3 The most famous quarry in the world: Carrara, Italy

No artist looms so large in Western consciousness and culture as Michelangelo Buonarroti, the most celebrated sculptor of all time. And no place on earth provides a stone so capable of simulating the warmth and vitality of human flesh and incarnating the genius of a Michelangelo as the statuario *of Carrara, the storied marble mecca at Tuscany's northwest corner.*

From E. Scigliano, *Michelangelo's Mountain: The Quest for Perfection in the Marble Quarries of Carrara* (2005)

We have already illustrated this chapter with a variety of quarries which produce building stones, and we have highlighted the high quality of Portland stone in the British context, but in the worldwide context nowhere exceeds the history and charisma of the Carrara marble quarries in the Apuane Alps, which are located along the Adriatic coast in Northern Italy—the most famous quarry complex of them all. The history of this quarrying site and Michelangelo's life are well covered in Scigliano (2005) and so we concentrate here on the quarrying method. (Note that later in the book, Chapter 8, we discuss the mechanical properties of Carrara marble in connection with the consequential 'bowing' of the marble slabs when used as decorative cladding for important buildings.)

Although the quarrying of the Carrara marble has been underway for more than 2500 years, its geological formation occurred long before that. This marble was formed by the metamorphism of 200-million-year-old Lower Jurassic limestones during the Alpine orogeny caused by the collision of the African and European tectonic plates which began in Late Cretaceous times some 65 million years ago. The limestone recrystallised in response to the increase in temperature and pressure linked to this tectonic event with the properties of the marble being determined by the composition of its parent limestone. The purest Carrara marble obtained from these Tuscan quarries was used in the carving of the renaissance masterpieces of Michelangelo. However, for sculptures destined to remain in the open air, particularly in cities where the erosive properties of polluted air need to be considered, another variety of marble containing some silica (quartz) is desirable even if more difficult to carve. An example of the use of the more silica rich variety of Carrara marble is the Parnassus frieze of the Albert Memorial discussed in Section 7.2. This variety is termed 'Campanella marble' because of its ability to ring like a bell when struck with a hammer, a property imparted by the silica enrichment.

One of the individual quarries within the whole Carrara quarry complex is shown in Figure 4.15(a) and the close-up in Figure 4.15(b) highlights the presence of natural fractures within the marble, the sides of which have become stained by natural water flow.

Figure 4.15(a) One of the individual quarries within the Cararra marble quarrying complex, Italy. Panel (a): (Photograph by the authors and courtesy of Imperial College Press, London).

Figure 4.15(b) A close-up of one of the benches in Figure 4.15(a), illustrating the presence of natural, water-stained fractures in the Carrara marble.

Figure 4.16 (a) The method of cutting Carrara marble using wire with diamond encrusted inserts. Panel (a): (Photograph by the authors and courtesy of Imperial College Press, London).

Figure 4.16 (b) A close-up of a diamond insert on the marble cutting wire.

The marble is cut by using wire which has diamond encrusted inserts at frequent intervals, see Figures 4.16(a) and (b). This method of *in situ* cutting and hence extraction of the marble is versatile and efficient. Because marble is a metamorphic rock which has been produced under conditions of high stress, it can deform when used for external cladding and then subjected to large temperature variations, see Section 8.6.

4.3.4 Preparation of building stones

As can be seen from the previous figures in this chapter, the types of rocks being quarried come from a variety of rock mass types, and in some cases have to be 'seasoned' after excavation, rather like timber. For example, Portland limestone following its excavation is relatively soft and easy to carve. It is therefore often excavated, carved into its required form and then left to season (harden) before being incorporated into a building. After excavation, any rock usually has to be sawn on site to the required shape, e.g., Figures 4.17, 4.18(a) and (b).

Sedimentary rocks are naturally fractured and easier to quarry than granite, so the rock blocks can sometimes be obtained more or less directly with roughly flat faces, Figure 4.19,

Figure 4.17 Sawing granite to the required dimensions.

(a)

(b)

Figure 4.18 Granite stone cut to the required shapes for (a) constructing the London cycleway and (b) setts to be used for upgrading a town square.

Figure 4.19 Use of irregularly shaped sedimentary building stone blocks in 'vernacular' style.

but for more prestigious buildings the blocks will have to be sawn into the precise dimensions required. In the case of a metamorphic stone like slate, roofing tiles can be split and shaped from the excavated rock by exploiting the intrinsic rock fabric, see Section 3.7.1.

Where the surface of the stone is an important feature of the building, the stone surface is dressed, i.e., prepared according to the architectural requirements. This was done with a hammer and chisel in earlier days, Figure 4.20(a) and nowadays by machine, Figure 4.20(b).

One particular form of texturing the building stone surface is 'vermiculation', as illustrated in Figure 4.21. It is used to make the plain surface more interesting and can be observed in different geometries as variations on the overall theme. The word 'vermiculation' is derived from the Latin *vermiculus* meaning 'little worm', with the pattern reminiscent of worm tracks in mud, and is instantly recognisable.

4.3.5 The largest quarried stone block

The size of quarried stone pieces is determined by two factors: the required size of the blocks and the capacity of the moving equipment. In Figure 4.22, we show a limestone block left by the Romans near a quarry at Baalbeck (Heliopolis—the city of the sun) in Lebanon. For unknown reasons, this stone, weighing ~1,240 tons, is known as the 'Stone of the Pregnant Woman' and may have been prepared for a nearby temple about a kilometre away for the god Jupiter. Its dimensions are about 20.5 m long, 4 m high and 4 m wide.

Since the Figure 4.22 photograph was taken in the 1960s, another large rectangular style quarried block, estimated to weigh about 1650 tonnes, was discovered in the same quarry

(a)

(b)

Figure 4.20 (a) Rough and (b) shaped decorative stone surfaces.

Figure 4.21 Vermiculated blocks outlining the archway at the entrance to Devonshire Square in London.

Figure 4.22 Large limestone block quarried by the Romans, Baalbeck, Lebanon.

and adjacent to the Figure 4.22 block. That one measures 19.6 m long, 6 m wide and at least 5.5 m high. One can only marvel at the imagination, tenacity and perseverance of the Roman quarrymen and builders in not only designing temples with such enormous stone components, but actually quarrying and shaping the blocks with just hand tools. It is a pity that in this case the difficulties of transporting such super-sized blocks to the temple site defeated them, or perhaps other factors were involved in causing the blocks to be left unused.

4.4 BUILDING STONE EXPOSURE, DETERIORATION, 'BACK TO THE EARTH'

In this short section, we briefly note the degradation of the stone itself, with further discussions in Sections 7.3.3 and 7.3.4 specifically related to the sandstone of Durham Cathedral, and more extensively in Chapter 8 on the overall deterioration of stone buildings. There is a variety of factors leading to the degradation of building stones relating to their physical and chemical characteristics, which in turn are a function of their geological origin. For example, and as we have noted in Chapter 3, essentially impermeable granite building stones are specifically used where exposure to water is a particular hazard, e.g., for sea walls, lighthouses and at the base of external pillars and walls. With other building stones, a main factor influencing degradation is the way in which the building stone bedding planes are orientated in the building, as indicated in Figure 4.23(a), where placing the stone with its bedding planes in the vertical direction leads over time to flaking at the exposed surface; however, when the bedding planes are horizontal, this type of degradation is much reduced—a fact recorded on page 140 of the 1806 *A Dictionary of Arts and Sciences, Vol. 1* by G. Gregory, see Figure 4.23(b). Note also in Figure 4.23(a) that the sub-vertical cracks in the central building stone has induced a crack in the building stones—both above and below.

In masonry, the stones must be so cut, as to lie in the same direction as they did in the quarry, in order that their strength and solidity may be fully employed; for if their position be changed, and they are placed vertically, they are apt to split; and the smallest crevice in the foundation will produce a great cleft in the superstructure of a building.

(a) (b)

Figure 4.23 (a) Building stone splitting along the bedding planes—western side entrance to the Bonsecours Market, Montréal, Canada. Note especially the cracking at the exposed right-hand surface of the building stone. (b) Salient extract from page 140 of the 1806 *A Dictionary of Arts and Sciences, Vol. I* by G. Gregory.

In addition to cracking caused by adversely loading a sedimentary building stone's structure, severe temperature fluctuations can damage the stone and lead progressively to major deterioration.

* * * * *

The subject of building stone deterioration is part of the wider subject of disintegration. As the Historic Scotland leaflet on *Masonry Decay* so succinctly phrases it: "All materials decay: some types of stone deteriorate faster than others. Softer sandstones decay faster than harder sandstones and they, in turn, decay faster than granite." In the geological context, Seward (1945) eloquently considers the overall subject as follows: "Rocks crumble into dust and from the dust new rocks are made; matter is indestructible; change and decay are stages in constructive processes. What is called destruction is not annihilation, it is an undoing of something which was put together, the conversion of a complex structure into its component parts." We will take up this theme again in Chapter 8 on the deterioration of building stones and stone buildings in the context of the inevitability of ruins.

NOTES

1 A touchstone is used to identify precious metals.
2 A freestone is a building stone that can be freely cut in any direction.

Chapter 5

Stone buildings—pillars, lighthouses, walls, arches, bridges, buttresses, roof vaults, castles, cathedrals and lettering

5.1 INTRODUCTION

Following Chapter 2 on the geological background to building stones, Chapter 3 on how to recognise the different types of building stone and Chapter 4 on the birth-to-death life of a building stone, we now concentrate on the structural components of stone buildings, starting with the simple geometries of a pillar, a lighthouse and a wall. The mystery of why stone arches can be much more stable than they may appear is explained with the three stone bridges across the tidal reach of the Thames as examples. We also explain how a network of stone arches in a roof (the vaulting) is based on an extension of the same single arch principle. These structural components have been used to construct stone castles and cathedrals. The chapter concludes with an explanation of decorative lettering in stone.

5.2 PILLARS, LIGHTHOUSES AND WALLS

5.2.1 Pillars

The simplest form of stone building is a pillar, i.e., one stone placed on top of another. However, even with this single architectural component, one is immediately faced with a series of potential problems, Figure 5.1. How high can the pillar be before the foundation stone fails because of the weight of the stones above? Is the ground strong enough to support a tall pillar? How long will it be before the building stone decays and the pillar collapses?

Calculation: Of the questions above, the one concerned with the height of the pillar and the strength of the lowest stone is the easiest to address: we just need to know the weight of the stone and its strength—or, more strictly, the density of the stone (weight per cubic metre) and its strength when loaded (how much compressive load per square metre it can sustain before breaking).

From these early data obtained by Watson (1911), we can create a graph of the mean density of 412 British building stones according to their geological age, see Figure 5.2. This has the satisfying trend that on average, the older the stone, the heavier it is likely to be. We now calculate as follows (using Watson's Imperial units, but with the SI units equivalents):

- Take a mean density of, say, 150 lbs/ft^3 ($\cong 150 \times 16.02$ kg/m^3).
- This exerts 150 lbs on each square foot of the base for each foot of pillar height (Fig. 5.1), or a bit more than 1 psi (*cf.* a typical car tyre pressure of 30 psi).
- Assume the compressive strength of the rock is, say, 500 tons/ft^2.
- Then the pillar height before the base block collapses is $500 \times 2240/150$ ft high.
- Which is 7467 ft $\cong$ 1.4 miles = 2.3 km.

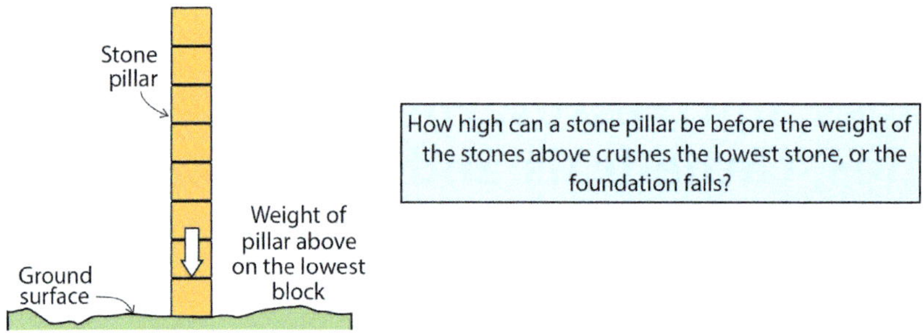

Figure 5.1 The strength of the building stones and the foundation in resisting the weight of the stones above.

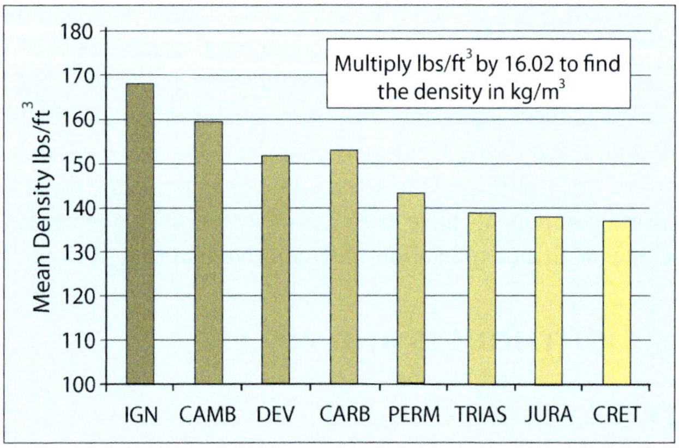

Figure 5.2 Summary of the mean density data in Imperial units according to the geological period for the 412 British building stones listed in Watson (1911). The abbreviations along the *x*-axis refer to Igneous and, older to younger, Cambrian, Devonian, Carboniferous, Permian, Triassic, Jurassic and Cretaceous.

By considering the result of this calculation and similar maximum pillar height values for weaker stones, we come to the conclusion that most building stones will be able to sustain very high pillar heights, much higher than conventional buildings or even skyscrapers. But there are other factors involved besides the stone strength. What would be the effect of weak ground below the foundation stone, e.g., as in the case of the clay and sand below the Leaning Tower of Pisa? Are any of the pillar stones severely misaligned or cracked? What effect does a strong wind have on an isolated pillar? What is the effect of a load being applied to the top of a pillar, perhaps not perfectly vertically applied, e.g., part of the roof of a cathedral?

The measurement of the mechanical characteristics of stone has a long history leading up to the sophisticated testing machines now available. We show part of a modern testing machine in Figure 5.3. In Figure 5.3(a), a cylindrical sample of stone is placed between two steel platens (loading cylinders) having the same diameter as the sample. The platens are then squeezed together, gradually increasing the load on the stone specimen until it breaks. The instrumentation around the broken sample in Figure 5.3(b) enables a record to be made of the amount of longitudinal compression and circumferential expansion that occurs—before, during and after the specimen failure. In this way, and with different types of loading, the complete mechanical behaviour of stone can be studied; in particular, the strength of a laboratory specimen or an *in situ* pillar can be increased by lateral confinement, Figure 5.3(c).

(a) Before the strength test

(b) After the strength test

(c)

Figure 5.3 (a) and (b) Testing the compressive strength of igneous rock at the SP Technical Research Institute, Borås, Sweden. (c) Enhancing the strength of a chalk pillar using metal restraining bands, Guildford, England.

Through this type of laboratory test and others, the overall characteristics of a building stone can be determined. For example, the annual Natural Stone Directory produced by the QMJ Group Ltd. in the UK provides key technical information for quarry stones listing colour, texture, porosity, absorption, density, slip resistance and abrasion resistance. Generally, the compressive, tensile and shear strengths of the stones are not specified because, as we have seen from the calculation above, the majority of stones are strong enough for conventional uses. Stone failure in a building is much more likely to occur because of pre-existing fractures in the stone, progressive flaking due to chemical effects, or being laid with the original bedding planes vertical instead of horizontal—rather than the stone being over-stressed because of the weight of the stones above. However, the circumstances can be 'interactively complicated' in some cases, e.g., the Leaning Tower of Pisa in Italy (183 ft/56 m high with its 5° tilt having moved the top horizontally by more than 16 ft/5 m). The relatively weak foundation material of clay and sand has enabled the tilt, but it has also dampened earthquake induced vibrations so that the tower has not yet been significantly affected by earthquake ground motion.

As an aside relating to present day techniques using modern materials, extraordinary tall buildings are now being built on New York's hard rock foundation, e.g., 432 Park Avenue in New York City. This building has 86 storeys and is 426.5 m high, the tallest residential tower in the Western Hemisphere, with an aspect ratio (height to width ratio) of 15:1. Being residential, it is important that the tower does not move too much in the wind, so the lateral acceleration has to be kept low. A nearby building, at 111 West 57th Street, being constructed also as a residential building, has an even larger aspect ratio of 24:1, will be 439 m high, and will be the world's most slender building. Counterintuitively, perhaps, modern skyscrapers are among the safest buildings because they have to be so carefully designed. In Great Britain, the tallest building is the Shard at 310 m high with 95 storeys, the foundations being built on layers of alluvium and permeable river gravels above the London Clay at a depth of 9 m.

Note that, as in the architectural example in Figure 5.4, stone pillars can support heavy weights—assuming that the foundations are adequate and that there are no significant

Figure 5.4 The façade of Covent Garden Royal Opera House in London.

sideways forces. A description of the pillar-type Monument in London (which commemorates the Great Fire of London in 1666) is included below.

The Monument (Fig. 5.5), a fluted Doric column, is one of the tallest isolated stone columns in the world. It is 202 ft (62 m) high and close to the bakery location in Pudding Lane where the Great Fire started on 2 September 1666, and on the site of the first church to be destroyed by the fire: St. Margaret's Fish Street. It is constructed with Portland stone, topped by a gold leafed 'flaming' urn symbolising the Fire, and has a stainless steel viewing cage near the top. Internally, the column has a 311 step cantilevered stone staircase. Originally, it had a caretaker and, in 1847, a resident mouse who reputedly published a book: *History of the Adventures of a Mouse Written by Himself.* The Monument was opened to the public (the City of London's first tourist attraction) in 1890, the cost of construction having been £13,450, ≅ £1.7 million in today's money. In 2007, there was a major restoration costing £4.5 million. It is well worth a visit.

The style of pillars in classical architecture was well defined in Ancient Greece and the Roman Empire, Figure 5.6, and these have been used extensively in Britain since the late 15th to early 17th century Renaissance period. There was particular emphasis on the use of square/rectangular forms in building façades together with the different types of pillars, i.e., Tuscan, Doric, Ionic, Roman and Corinthian, of which Doric, Ionic and Corinthian are the most widely used in British architecture, e.g., the Corinthian type in Figure 5.4 at the Covent Garden Royal Opera House in London. Note that the type of pillar can be quickly recognised

Figure 5.5 The Monument to the Great Fire of London.

Figure 5.6 Doric, Ionic and Corinthian are the three types of classical pillar most widely used in British architecture.

Figure 5.7 Illustrating the three main styles of the stone pillar capitals in a 'ruined format': left, Corinthian; centre, Doric; right, Ionic (upside-down). (After an illustration by J. Gwilt from *A Treatise on the Decorative Part of Civil Architecture* by W. Chambers)

in the first instance by the pillar capital style, i.e., whether there is a plate (Doric), a scroll (Ionic) or scrolls and acanthus leaves (Corinthian) at the top of the pillar, remembering that other pillar components contribute to each style. These types of pillar are further illustrated in an interesting 'ruined style' in Figure 5.7, which emphasises the different pillar capital styles. Architectural styles are discussed further in Chapter 6. The exterior pillars of Rylands Library in Manchester are shown in Figure 5.8.

Figure 5.8 Corinthian style pillar capitals in the foreground; Tuscan (similar to Doric) style pillar capitals above in the background. Rylands Library, Manchester Central Library.

5.2.2 Lighthouses

In the afternoon of time, A strenuous family dusted from its hands
The sand of granite, and beholding far,
Along the sounding coasts its pyramids and tall memorials catch the dying sun.
Robert Louis Stevenson (1850–94)

Many lighthouses, which are essentially hollow pillars, have exciting histories and, although often on rock foundations (rather than the London clay beneath the Monument pillar in London as described above), can be subject to extremes of weather because of their locations. Many of the lighthouses around the British coast, in fact an extraordinary number, 83, were built by six members of the Stevenson family, Figure 5.9.

The family tree in Figure 5.9 starts with Thomas Smith who was the stepfather of Robert Stevenson, having married Robert's mother, Jean Lillie (who was the widow of the first Alan Stevenson, 1752–1774). Thomas Smith was the first engineer appointed to the Northern Lighthouse Board soon after its inception and was already an experienced engineer. He educated and inspired Robert Stevenson who became the 'head of the Stevenson lighthouse building family' which consisted of his three sons, Alan, David and Thomas, and two grandsons, David and Charles. Robert Louis Stevenson, son of Thomas Stevenson, was originally

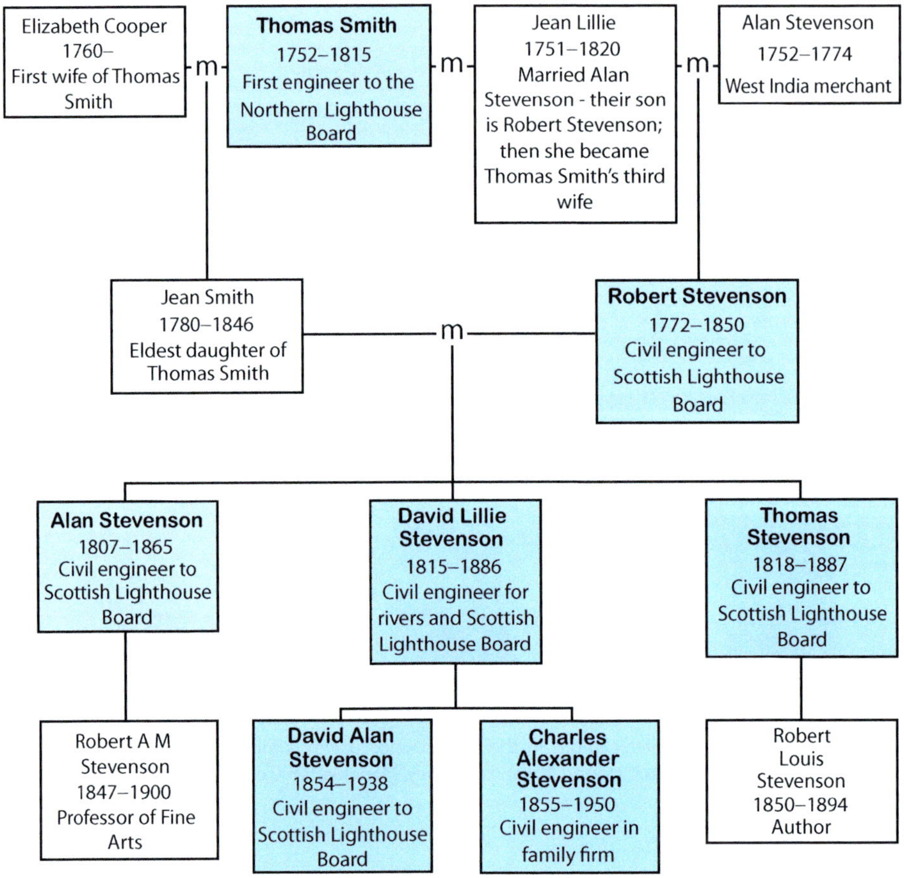

Figure 5.9 "The Lighthouse Stevensons"—family tree showing the male members of the Stevenson lighthouse building family with the engineers highlighted. Six members of the Stevenson family (i.e., excluding Thomas Smith) built 83 lighthouses around the coast of Britain. The horizontal lines containing a small 'm' indicate a marriage. The engineers are highlighted in blue. Note the influence of Thomas Smith: as stepfather to Robert Stevenson (1772), he encouraged Robert's engineering education and the two of them were the founders of the business. Robert's son, Alan Stevenson, was the designer and builder of the Skerryvore lighthouse featured in this chapter.

trained as an engineer but did not follow in the family's lighthouse business and became famous as an author, e.g., *Treasure Island* and *The Strange Case of Dr Jekyll and Mr Hyde.*

The construction period and number of lighthouses associated with the family members are as follows:[1]

- Robert Stevenson (1811–1833) — 17 lighthouses
- Alan Stevenson (1833–1853) — 12 lighthouses
- David L. and Thomas Stevenson (1854–1877), brothers — 28 lighthouses
- David A. and Charles A. Stevenson (1855–1937), brothers — 26 lighthouses

Alan Stevenson, the son of Robert Stevenson, made a detailed overall building plan for the Skerryvore lighthouse showing the precision required for the preparation of the individual granite blocks for each of the 97 courses (Fig. 5.10). The location of the lighthouse off the west coast of Scotland in shown in Figure 5.11.

To illustrate some of the problems encountered when building a lighthouse on a rock reef in the sea, the preparation of the constitutive igneous rock blocks and the construction

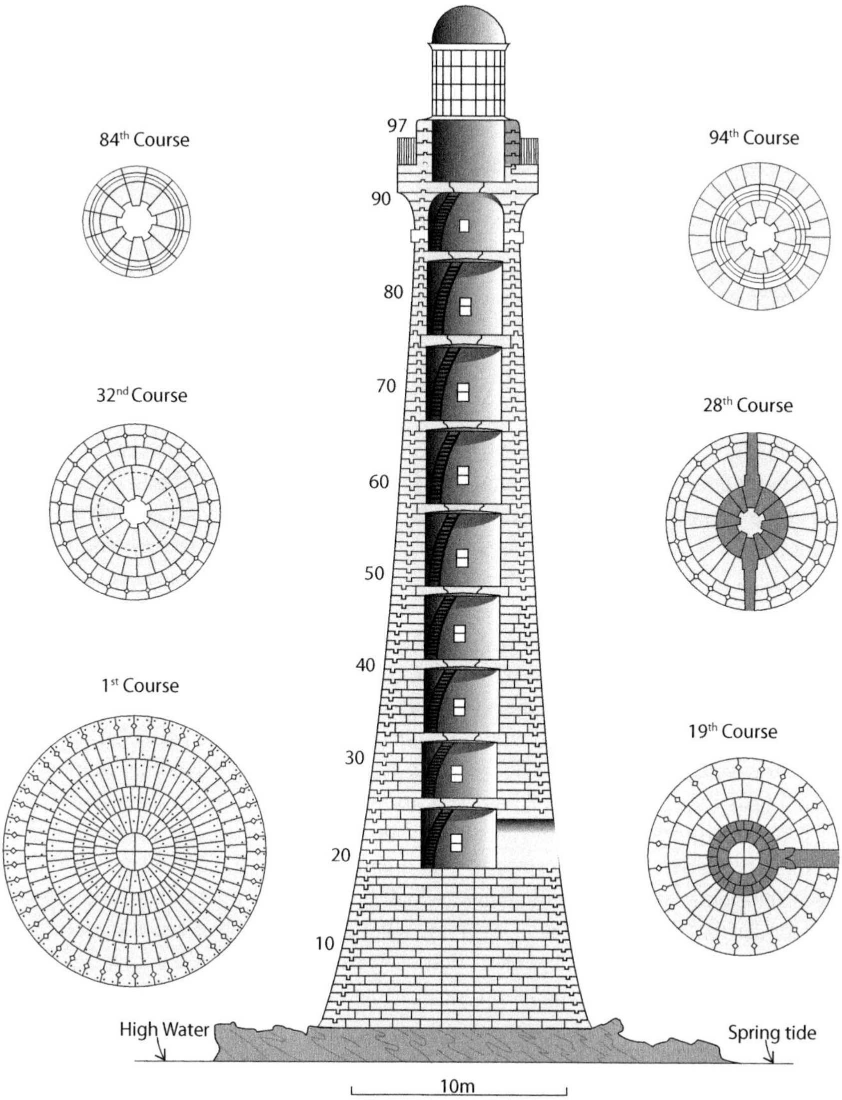

Figure 5.10 The Skerryvore Lighthouse with its 97 stone courses, located off the west coast of Scotland, designed and built by Alan Stevenson in the 19th century (diagram after Stevenson, 1848). The cross-sections of some of the courses indicate the huge amount of intricate work involved in hand preparing the individual stones many of which interlock with the underlying level and have somewhat different geometries at each lighthouse level.

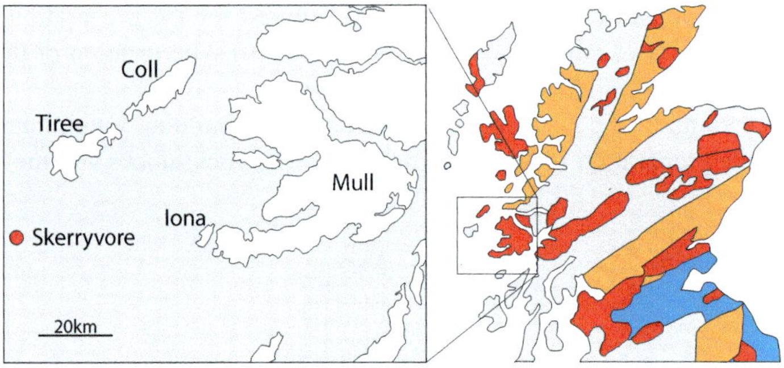

Figure 5.11 The Skerryvore lighthouse stands on the Skerryvore 'reef', a semi-submerged ridge of rock made of Lewisian gneiss situated 19 km SW of Tiree and almost due west of the island Iona. About Iona, Stevenson noted, "There is good reason also for concluding that the Mull stone (i.e., the Ross of Mull granite) is sufficiently durable, because it contains but a small proportion of micaceous matter, and in its texture closely resembles some of the blocks of St. Oran's chapel in the neighbouring Island of Iona, which have resisted the action of the weather, it is believed, for more than 600 years and still retain the marks left by the tools of the workmen."

and the stone block architecture of the Skerryvore Lighthouse as built by Alan Stevenson, we have included the following text extracts from his extraordinary 1848 book titled *Account of the Skerryvore Lighthouse with Notes on the Illumination of Lighthouses*. This book can be downloaded in PDF form at no cost—see the reference later in this section and the entry in the References and Bibliography section. In his preface to the book, Stevenson notes that,

> Although, in the course of the Narrative, I have occasionally noticed some special deliverances from danger, I have altogether neglected to record the remarkable fact, that, amidst our almost daily perils, during six seasons on the Skerryvore Rock, there was no loss of either life or limb amongst us. Those who best know the nature of the service in which we were engaged,—the daily jeopardy connected with landing weighty materials in a heavy surf and transporting the workmen in boats through a boisterous sea, the risks to so many men, involved in mining the foundations of the Tower in a space so limited, and above all, the destruction, in a single night by the violence of the waves, of our temporary barrack on the Rock, which had cost the toils of a whole season, will not wonder that I am anxious to express, what I know to have been a general feeling amongst those engaged in the work—that of heartfelt thankfulness to Almighty God for merciful preservation in danger, and for the final success which terminated our arduous and protracted labours.

At the beginning of Chapter 1, he introduces the nature of the problem:

"From the great difficulty of access to the inhospitable rock of Skerryvore, which is exposed to the full fury of the Atlantic, and is surrounded by an almost perpetual surf, the erection of a Light Tower on its small and rugged surface has always been regarded as an undertaking of the most formidable kind." He describes the site as follows: "The surface of the main or principal rock, on which the Lighthouse has been placed, measures, at the lowest tides, about 280 feet square. It is extremely irregular, and is intersected by many gullies or fissures, of considerable breadth, and of unlooked for depth, and which leave it solid only to the extent of 160 feet by 70 feet. The extremity of one of these gullies, at the south-east corner of the rock, forms the landing-creek, which is a narrow track of 30 feet wide, having deep water; and, with the help of some artificial clearing and dressing, which was executed with much difficulty, by blasting under water, while the other works were in progress, its sides and bottom are now comparatively smooth. At this place a landing can often be effected when the rock is unapproachable from any other quarter, although great inconvenience is felt from the surge, which finds its way from the opposite side of the rock, through the westward opening of the gulley in which the landing place is situated."

Describing the rock he says, "The rocks of Skerryvore have the same characteristics as those of the neighbourhood of Tyree, being what we may, perhaps, call a syenitic gneiss, as it consists of quartz, felspar, hornblende, and also mica. It will be seen, from the narrative of the progress of the works, that this rock was, from its hardness, exceedingly difficult and tedious to excavate." Following a discussion on the factors governing the overall shape and size of a lighthouse, Stevenson decided on the height of the pillar as 138.5 ft, and the radius of the base, at the level of about 4 ft above high water, as 21 ft. Then he had to decide on the curvature of the external wall, considering "Parabola, Logarithmic, Hyperbola, and Conchoid" mathematical shapes, deciding finally on the rectangular hyperbola. This, in its simplest form, is mathematically represented as xy = constant (note that as x decreases, so y increases), part of this type of curve being the curve of the lighthouse exterior, Figure 5.12.

The extreme difficulty of working in such turbulent weather is exemplified by the following quotation describing the time after the office design work and then the on-site work in the summer of 1838 in building a temporary elevated wooden housing for the workers and equipment, "on the 12th November, I received from Mr Hogben, the clerk and store-keeper at Tyree, the unwelcome intelligence that the Barrack-house had been destroyed, as was supposed, by the heavy sea of the 3rd November."

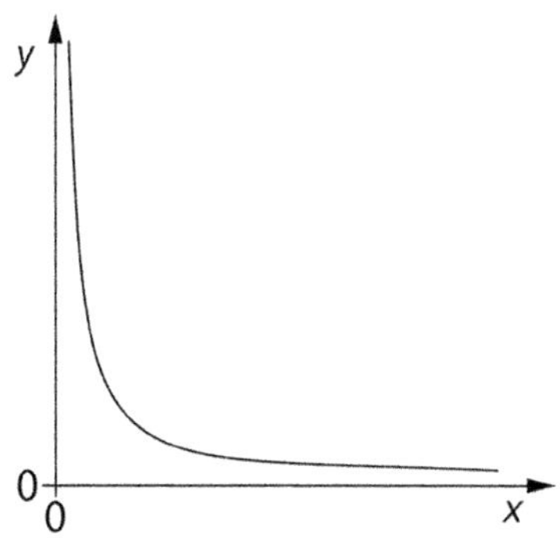

Figure 5.12 The shape of a rectangular hyperbola, *xy* equals a constant, plotted using *x,y* axes. The curve of the lighthouse wall from stone course 1 to 86 (see Fig. 5.10) is based on the upper part of this type of curve.

In addition to the Barrack-house problem, Stevenson also had a problem with the properties of the Lewisian gneiss from the Isle of Tiree. (Note that this rock type on the Isle of Lewis was introduced in Section 2.9.1.) After visiting various quarries, he decided to change to granite, noting that:

> Granite, indeed, is a material in many respects superior to sandstone, gneiss or porphyry. The first it greatly excels in durability and weight; and, as a stone for the work-yard, it is superior to the other two, from its property of being fissile, or easily split in any direction. In this respect it resembles certain parts of some sandstone strata which are commonly called liver rock, of which Craigleith quarry, near Edinburgh, furnishes an excellent example. Porphyry, and, I think, all other igneous rocks (excepting granite), gneiss also, and most of the other primary rocks, have not this property, being fissile only in one plane, so that quarries of those rocks generally turn out very uncouth or irregular stones, which, though they may in some favourable cases possess good natural beds, will always be found to have ragged and irregular joints, which, for the most part, are incapable of being properly dressed.

We noted in Section 3.2 that granite has excellent water resistant properties, but this fissile aspect discussed by Stevenson is an extra helpful property.

Changing the source of the stone to the island of Mull, Stevenson remarked that:

> "I know of no instance of a quarry so fully answering the most sanguine expectations as that of the Ross of Mull; and I have never seen a granite quarry of equally great resources, as regards both the quantity and the quality of the material produced." He goes on to explain the production of the rock blocks as follows, "When a mass of rock had been thus removed [by blasting], it was cut up into various portions by means of wedges, and finally subdivided into blocks, hammer (or as it is called quarry) dressed, according to rough moulds, whose dimensions exceeded those of the stones of the various courses of the building by a quantity which was considered sufficient to cover any casualty in the final dressing of the block." [In other words, the final shaping of the blocks was left until later.] He notes that, "The stone of the Mull quarries is a reddish or flesh-coloured granite, in which felspar predominates. About 13.66 cubic feet weigh a ton, and it is not quite so hard as the granite of Aberdeen [which we illustrated earlier in the book, Sections 2.9.2 and 3.2.3]."

Showing considerable understanding, Stevenson outlines in some detail the granite quarrying procedures, noting at one point that, "In selecting the position of the bore [i.e., blasthole], the direction of the seams and veins of the rock must be duly considered, with a view to employ the force of the explosion to the greatest advantage in separating the natural joints or beds of the rock, instead of shattering the solid masses or posts (as they are called in the language of the quarry) into shivers or fragments." He also describes in detail the later final finishing of the stones to the exact required dimensions. Note that in addition to its strength, density and resistance to water and weathering, making the Ross of Mull granite an ideal building stone for the lighthouse, its attractive pink-red colour and its ability to take a fine polish also make it an excellent decorative stone. Indeed, it has been used to great effect for the main cluster pillars that support the canopy of the Albert Memorial in London, shown in Figure 7.12 and described in detail in Section 7.2.

The Skerryvore on-site difficulties encountered by Stevenson when he began work on the rock can be summarised by his sentences:

> A more unpromising prospect of success in any work than that which presented itself at the commencement of our labours, I can scarcely conceive. The great irregularity of the surface, and the extraordinary hardness and unworkable nature of the material, together with the want of room on the Rock, greatly added to the other difficulties and delays, which could not fail, even under the most favourable circumstances, to attend the excavation of a foundation-pit on a rock at the distance of 12 miles from the land. The Rock, as already noticed, is a hard and tough gneiss, and required the expenditure of about four times as much labour and steel for boring as are generally consumed in boring the Aberdeenshire granite.

Stevenson's subsequent adventures in building the lighthouse can be read about in his book, which is referenced in the next paragraph.

In conclusion, the lighthouse is 137 ft high, weighs 4,300 tons, the walls at the base being 9.5 ft thick. After the seven-year construction time, the cost had been £90,000 pounds sterling, about £9 million today. The Institute (now Institution) of Civil Engineers described the lighthouse as "the finest combination of mass with elegance to be met with in architectural or engineering structures." The quotations in the previous paragraphs are from Alan Stevenson's book describing the building of this lighthouse: *Account of the Skerryvore Lighthouse with Notes on the Illumination of Lighthouses*, published by Adam and Charles Black Ltd., Edinburgh, 1848, which readers are encouraged to read at https://archive.org/stream/accountofskerryv1848stev#page/n9/mode/2up. The overall story of *The Lighthouse Stevensons* has been compiled by B. Bathurst (1999).

5.2.3 Walls

If a pillar is extended sideways, a wall is created—which can be built as a dry stone wall or a mortared wall. Robert Frost remarked that before he built a wall he would ask to know what he was walling in or walling out: the 'walling in' could refer to dry stone walls used to contain livestock, the mortared walls of houses for protection from the elements, or prisons for confining the inmates. The 'walling out' could refer to single long walls like the Great Wall of China (built in the 3rd century BC) and Hadrian's Wall (built during the Roman occupation) across the north of Britain, or wall enclosures around towns or castles for protection from attackers. For important structures, stone walls replaced timber walls in Britain during Roman times in the reign of Trajan who is famous for his Trajan's triumphal column in Rome, see the wall building scene in Figure 5.13, which shows well prepared ashlar (square cut and finely dressed masonry with thin mortaring). Also, the earlier earthen forts were replaced by the Romans using stone and those at York and Chester had stone-revetted (i.e., masonry faced) ramparts and stone internal buildings.

The most famous Roman stone wall in Britain is Hadrian's Wall (a World Heritage Site) which links Wallsend on the east coast to Bowness-on-Solway on the west coast and was the northern frontier of the Roman Empire for nearly 300 years, following its construction from 122 AD onwards. The facing stones of the wall, 73 miles long, are mainly sandstone but the wall is partly built on and with the Whin Sill stone, which is a very hard dolerite (noting

Figure 5.13 Roman wall building, portion of Trajan's column. Note the brick bonding type used here by the Romans is known now as 'stretcher bond' (Fig. 3.101) and is widely used in Britain.

that in Emily Brontë's novel *Wuthering Heights*, Heathcliff is described as "Rough as a saw-edge, and hard as whinstone!").

Naturally, such ancient walls as this have been subject to the ravages of time and in Rome itself, the 8-mile city wall, built by the Emperor Aurelian in AD 271 and which is Rome's biggest monument, is no exception and in parts is in danger of collapsing, e.g., a 14 m section close to the British Embassy gave way recently following seismic disturbances. In Roman territories subject to earthquakes, such as at the Dougga site in Tunisia, the walls are known to have been reinforced by periodically using larger stones within the masonry in addition to heavy quoins at the edges, see Figure 5.14. This area in the Atlas mountains is subject to earthquakes because it lies at the junction of two tectonic plates: the Eurasian plate to the north and the African plate to the south (see Fig. 2.6).

In London, a significant Roman legacy is the remainder of the London Wall, which originally circumscribed the first major settlement by the river Thames, and which now outlines the modern City of London (Fig. 5.26). After the initial wall, a new riverside wall was built between 240 and 360 AD but, since AD 410 when the Romans left Britain, there have been many alterations made by successive invaders and inhabitants. What now remains of the Roman Wall, although significantly restored in places, still indicates how it was made, much of it with Kentish ragstone (the only hard rock available in the south-east of England, see Section 3.4 on limestones) interleaved with thin red tile bonding courses, Figure 5.15(a) and (b). Many Roman artefacts associated with the Wall can be found in the Museum of London.

Figure 5.14 Roman wall internally reinforced with larger building stones, rather like internal quoins, at Dougga, a World Heritage Site in northern Tunisia, which is subject to earthquakes.

(a) 

Figure 5.15 Portions of the Roman London Wall, illustrating the Kentish ragstone building stone with red tile bonding courses.

(b) 

Figure 5.15 (Continued)

In the countryside, dry stone walls are an attractive feature and are made from local stone, usually from stone easily available at the ground surface and minimally prepared, so these walls are a good indication of the type of stone that is locally available. In Figures 5.16–5.20, we illustrate how the geological nature of the stone plus the weathering effects influences the appearance of such dry stone walls and old house walls—leading to a 'vernacular architecture' as determined by easily available local stones. These examples were photographed at the National Stone Centre in Derbyshire near the Peak District National Park, which is well worth a visit and where examples of dry stone walls using stones from different British locations have been constructed and can be viewed. The walls are presented in geological order—from 'oldest to youngest'.

The Dry Stone Walling Association held its 50th anniversary meeting in May 2018, in Kirkby Lonsdale, Cumbria, England. The award for Best Master Craftsman Waller was won by the aptly named Andrew Mason. Readers interested in dry stone walling can find further information at www.dsw.org.uk.

Gooley (2015) makes some interesting notes about dry stone walls. Firstly, the stone is most likely to be locally sourced and hence is a direct indicator of the underlying geology.

Figure 5.16 Dry stone wall using Cumbrian green slate from the Ordovician Borrowdale Volcanic series, England.

Figure 5.17 Dry stone wall using glacial erratic, Lower Devonian, granite boulders, Mid-Galloway, Dumfries and Galloway region in Scotland.

Figure 5.18 Dry stone wall constructed with Middle Devonian, Caithness Flagstone (siltstones and fine-grained sandstones) which was also used for the construction of ancient houses at the Skara Brae site in Orkney.

Figure 5.19 Dry stone wall using Black Whin stone, a quartz dolerite stone (a coarse-grained variety of basalt, see Fig. 2.4) found in southern Scotland and northern England. The angularity and hardness of the stone discourages trimming of the individual blocks and, as mentioned earlier in the text, stimulated Emily Brontë in her *Wuthering Heights* novel to describe Heathcliff as "Rough as a saw-edge, and hard as whinstone!"

Figure 5.20 Dry stone wall constructed using Middle Jurassic, Great Oolite limestone from the Jurassic stone belt, see Figure 2.11.

Secondly, if the stones are dark they are probably acidic stone but, if they are light in colour, they are probably limestone which is alkaline. In Britain, the use of dry stone walls is generally restricted to boundary walls and small farm buildings because there are limits to the height of unmortared walls.

Another interesting factor to note is the surface of building stones. If the surface has not been 'dressed', i.e., chisel tooled to improve its finish, the surface may exhibit *fractographic*[2] patterns where the stone has been broken—either naturally, in the quarry or via the mason. In Chapter 2, we discussed the natural stresses which are present in the Earth's crust, both vertically from the weight of the rocks above and horizontally from the forces exerted by tectonic plate movement. These stresses cause fracturing in the natural rock which we highlighted in Chapter 4 in the context of quarrying. But whether the rock is fractured naturally or via the mason's hammer, fractographic lines can be produced and often identified in stone walls, Figures 5.21(a), (b) and (c).

The pattern of lines radiating from the impact point/splitting point, i.e., the point of origin of the fracture, Figure 5.21(b), (c) and (d), are referred to as plumose structures because of their resemblance to feathers (from the Latin *pluma*, a feather). The lines are microscopic steps on the fracture surface and they reveal the direction of propagation of the fracture.

By drawing lines at right angles to these, the shape of the fracture front is revealed as it propagated away from the origin, as can be seen in Figure 5.21(c) and (d). Sometimes small ridges which are parallel to the fracture front can be found on the surface: these are termed arrest lines as they record the position where the fracture temporarily stopped growing, Figure 5.21(d). Plumose patterns are delicate features and characteristic of natural extensional fractures (joints). Such features would be destroyed by the fracture-parallel slip which occurs on natural shear fractures (faults).

Rock masses are often intensely fractured near the Earth's surface but the fractures tend to be planar at the scale of a building stone, thus providing a convenient overall flat surface

Figure 5.21(a) Part of the exterior wall of Holy Trinity Parish Church, Ilfracombe, England. Note the fractographic lines in the large stone at the left, lower centre of the photograph, see Figure 5.21(b).

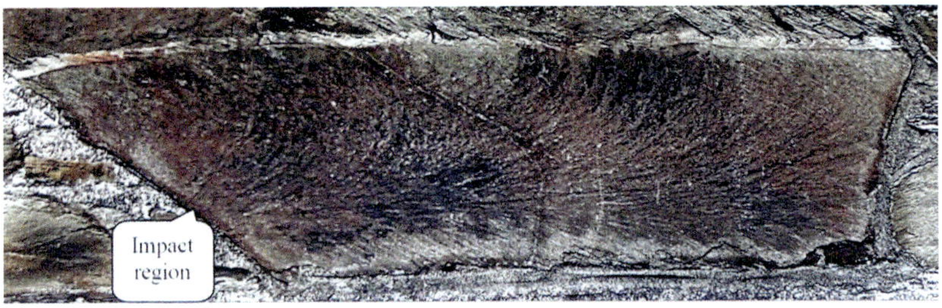

Figure 5.21(b) Magnified image of the large lower stone in Figure 5.21(a) showing the region of impact and the fractographic lines across the whole stone fracture surface.

Figure 5.21(c) Another example of a building stone fractured by the mason's hammer, Ilfracombe, England. Note the fractographic surface texture.

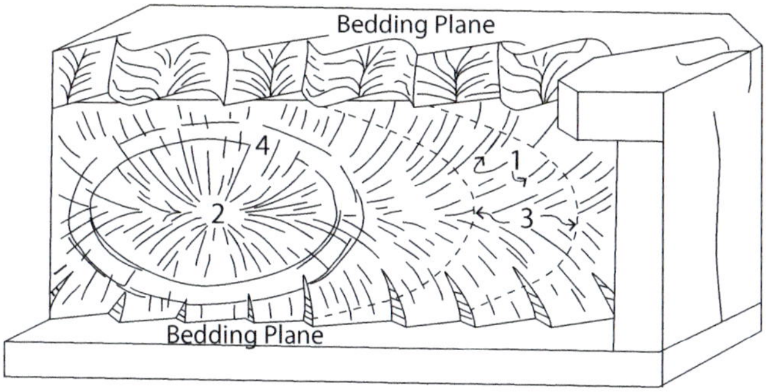

Figure 5.21(d) The geometry of fractographic lines on a vertical joint between two bedding planes. The lines denoted by 1 are the plumose lines, 2 denotes the point of origin of the fracture, the dotted lines 3 are drawn at right angles to the plumose lines and reveal the shape of the fracture front as it travels through the rock and 4 are arrest ridges, small ridges on the fracture surface that show the positions where the fracture temporarily stopped during its propagation.

when they are combined into the facing of a wall. Also, more often than not, over geological time groundwater will have permeated through the fractures in the rock mass and stained the fracture surfaces, usually a yellowish-brown because of the ubiquitous iron oxide content. This means that the attractive stone wall in Figure 5.22 actually represents a reconfiguration

Figure 5.22 Wall of natural rock fracture surfaces, Quayfield Road, Ilfracombe, England.

of pieces of natural rock fractures where none of the block faces have been artificially formed or dressed by a mason.

We discuss castles further on in this chapter (Section 5.4), but pause here to look at one portion of the exterior curtain wall of Cardiff Castle, Figure 5.23, with its three different types of limestone and the reddish course of Radyr breccia (which was introduced in Section 3.8.1) and which caps the remains of the original Roman fort wall.

Examples of the style of walls built with locally sourced stone are shown in Figure 5.24(a) and (b). A portion of the Telfer Wall in Edinburgh is shown in Figure 5.24(a) and is part of the remains of one of Edinburgh's late medieval town walls, dating back to ~1630. It is named after its mason, John Taillefer. The wall in Figure 5.24(b) is the base of the castle wall in Guildford, the stone components being (bottom to top) chalk, sandstone and flint.

Referring back to Robert Frost's remark at the beginning of this section, that before he built a wall he would ask to know what he was walling in or walling out, we note that a single, continuous wall is used to indicate a major territorial boundary, such as the wall we mentioned earlier, the Great Wall of China, which is ~4540 miles (7300 km) long, stretches from the Gobi Desert to the Jieshishan Mountain in Korea, dates back to 700 BC, contains billions of bricks and is the largest masonry project ever undertaken, and the largest artificial structure of any kind in the world. Although the earthen parts of the wall are eroding away, the stone sections remain, albeit in various states of preservation and degradation, Figure 5.25.

Figure 5.23 Portion of the exterior of Cardiff castle wall. Pink Carboniferous limestone to the left, dark Carboniferous limestone to the right, Lias limestone along the base (part of the old Roman wall), with a single course of red Radyr stone above it (the seventh course above the grass).

In Britain, the 73-mile Hadrian's Wall across England was built to indicate the most northern extent of the Roman Empire. These long, continuous walls not only indicated boundary locations but were also defensive structures; however, if the area to be defended was a much smaller, local area, then an irregular, rectangular or circular area was walled around, as in city walls. The rectangular wall geometry is typical of the Roman towns' defensive walls, as illustrated by the outlines of the main Roman towns in England, plus the more

(a) Telfer Wall, Edinburgh

(b) Base of castle wall, Guildford

Figure 5.24 Examples of local walls (see text).

Figure 5.25 Well preserved section of the Great Wall of China (near Beijing).

detailed plan of London, Figure 5.26(a), and Roman Chester in the north-west of England, Figure 5.26(b). Note the close up of one of the original Chester Roman building stones in Figure 5.27.

There is an additional important point about the stability of stone walls at their right-angled corners: some form of strengthening is required to maintain their structural integrity,

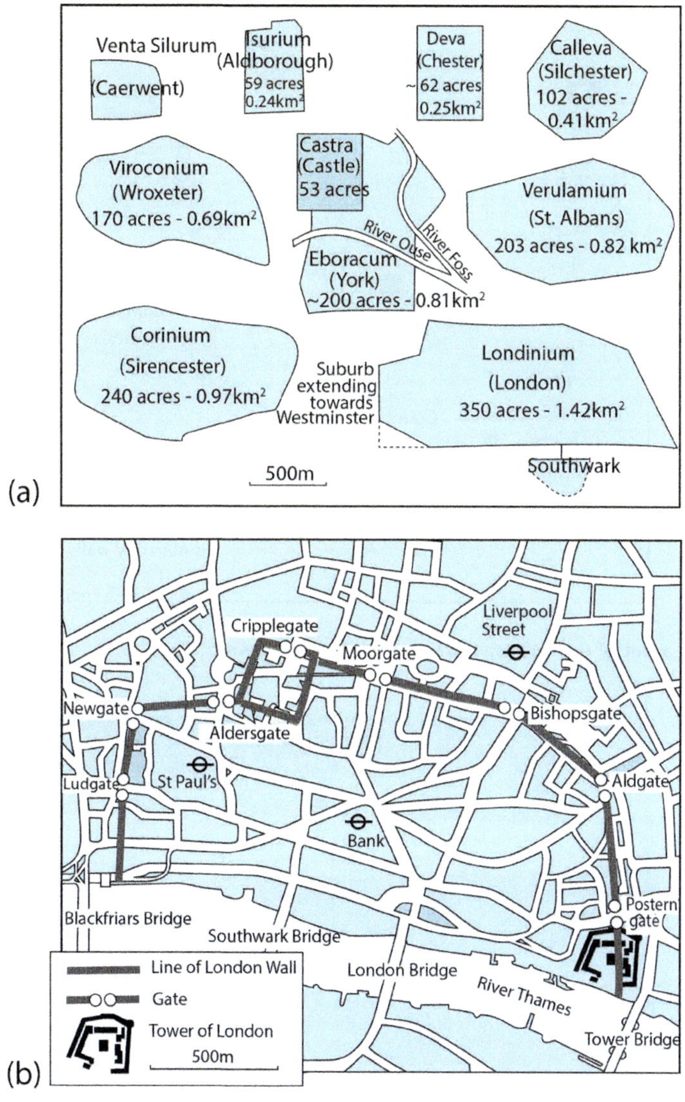

Figure 5.26(a) Upper: geometries of the Roman walls surrounding early towns in England. Lower: plan of the Roman London Wall. Note the number of locations ending in 'gate'—going counter-clockwise from the Tower of London at the bottom right, the Wall passes Postern-gate, Aldgate, Bishopsgate, Moorgate, Cripplegate, Aldersgate, Newgate and Ludgate. (Redrawn from Harris, E. (2009) *Walking London Wall*. Stroud, UK: The History Press)

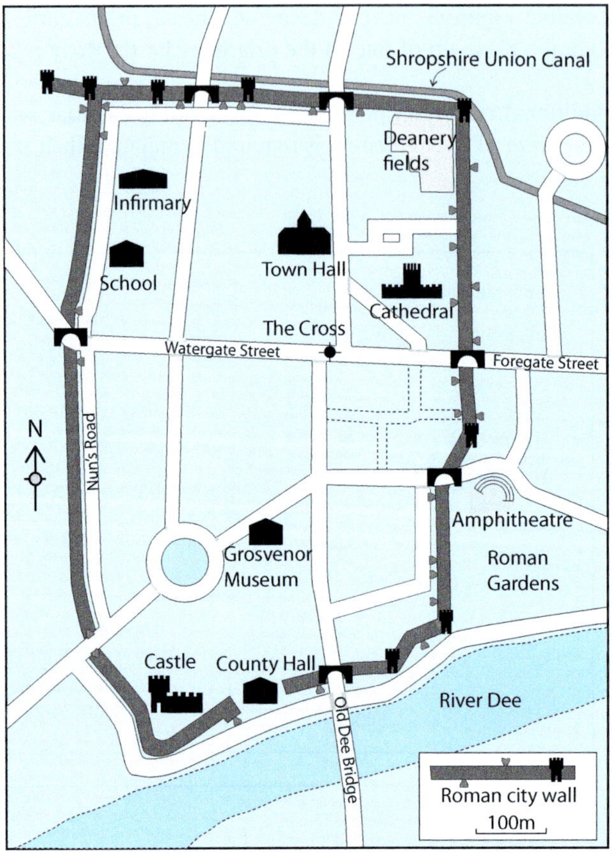

Figure 5.26(b) Layout of the defensive walls of the Roman town of Chester.

Figure 5.27 Part of the original Chester City sandstone Roman Wall.

Caerphilly Castle Newark Castle Colchester Castle

Figure 5.28 The use of quoins to strengthen the corners of stone buildings.

i.e., use of quoins. This word, pronounced like 'coin', is an old French word meaning 'corner'. The quoins are used to join two walls meeting at a right angle and hence enhance the building's strength, but they have also become a decorative aspect of a building's architecture, providing both a visual frame and additional interest, Figure 5.28, so much so that they are often included in modern buildings as architectural visual extras even when they are unnecessary.

Note how effective the quoins have been in the Newark Castle example in Figure 5.28: whilst the walls themselves are severely degraded, the quoins are 'holding the building together', and provide strengthening in the Colchester Castle example, although the upper ones have been replaced. Note that Colchester was one of the first towns in Britain.

Another structural aspect of stone buildings is the openings for windows and doors and the need to ensure their stability. As well as using stone for lintels over a window or door aperture to support the wall, they are often completely framed in stone. In addition to being aesthetically pleasing, this further stabilises them. The exterior of Salisbury Museum shown in Figure 5.29 illustrates both the use of stone around the window opening plus the opportunistic use of garnered stones during the construction of the original building. In the next section, we extend this discussion of walls to arches and roof vaults.

5.3 ARCHES, BUTTRESSES, BRIDGES AND ROOF VAULTS

Following the discussions in the previous section on pillars and walls, we now consider further components of stone buildings—i.e., the arches, buttresses and roof vaults. There is an interesting mechanical explanation of why stone arches are much more stable than they may appear and, as examples, we illustrate three low arch bridges across the Thames. The internal arches in stone buildings are stabilised by external buttresses, and roof vaults have a variety of designs based on both structural and decorative factors. These are the subjects that provide an understanding of the reasons why major stone buildings such as cathedrals have

Figure 5.29 The exterior of Salisbury Museum.

their characteristic architectures and stability—noting that we will discuss the evolution of specific architectural styles in Chapter 6.

5.3.1 Arches

A slab of stone can be used as a lintel to ensure stability above a wall opening, e.g., above a window or door. However, in order to achieve the same stability for larger openings such as bridges or cathedral naves and windows, it is not practical to use a single stone, so the multi-stone arch was developed, Figure 5.30. In the case of the Saxon doorway, Figure 5.30(a), it is easy to see intuitively why the reused Roman tiles in the triangular shape are stable, but in the case of the earlier semi-circular arch in Figure 5.30(b) the stability is not so self-apparent, especially the fact that it has remained standing for hundreds of years. Why do the central stones in the arch not just fall out?

Looking closely at the stones around the arch shown in Figure 5.30(b), it can be seen that they are not exactly rectangular in cross-section; rather they are slightly wedge-shaped, thus enabling them to be formed into a semi-circle. This is known as a Voussoir arch type of geometry, Figure 5.31.

When the individual stone blocks in an arch are of the Voussoir type (i.e., arranged as most of a semi-circle), Figures 5.30(b) and 5.31, the individual stones, the Voussoir blocks, have sides which are all directed towards the centre point. In this way, the line of force

(a) 'Arrowhead' doorway in the oldest surviving church and the only Saxon building still standing in Colchester.

(b) Newport Arch, a 3rd century limestone arch, the north gate of the Roman city wall.

Figure 5.30 Arches from Saxon and Roman times. (a)'Arrowhead' doorway with reused Roman tiles in the tower of Holy Trinity Church, Colchester, *ca.* 1000 AD. (b) Newport Arch, a 3rd century limestone Roman arch still standing in Lincoln, England, and still being used by traffic.

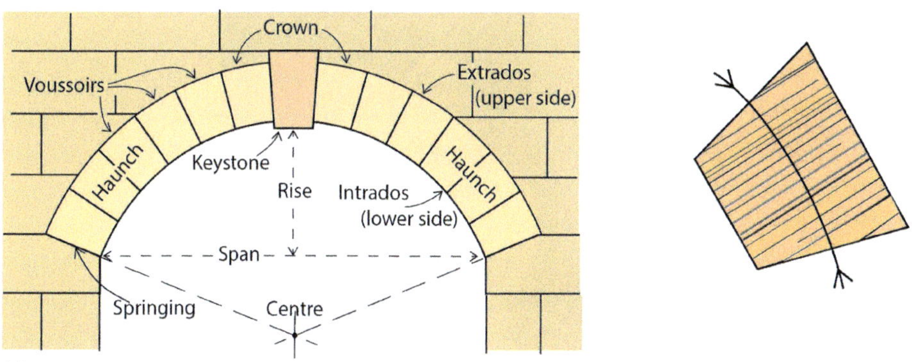

(a) (b)

Figure 5.31 (a) Sketch of the Voussoir arch components. (b) An individual Voussoir block with the arc of compression perpendicular to the upper and lower sides of the block. The Voussoir stone blocks should be cut so that the compression is at a high angle to the bedding.

induced by the weight of the stones is directed around the arch. When, a sedimentary stone is used, it is best if the line of force in each block acts perpendicular to the main bedding planes to avoid splitting of the blocks, see Figure 4.23. Since all the Voussoir blocks in the arch have the same geometry, they should all have the bedding planes orientated as indicated in the sketch in Figure 5.31(b).

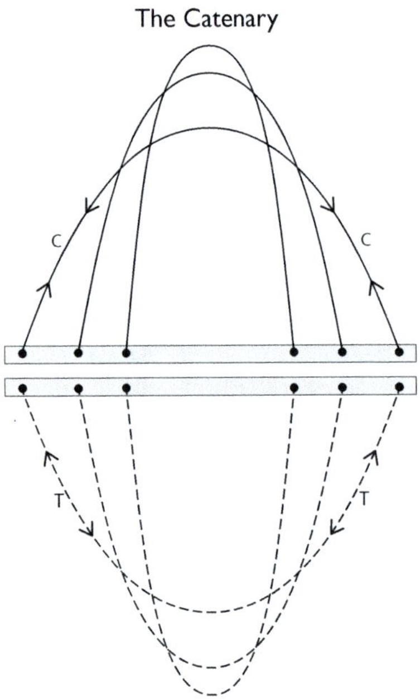

Figure 5.32 The catenary arch—formed by the inversion of a hanging string/chain.

Compression Tension Shear

The explanation above of the arch with a profile geometry which forms part of a circle is satisfying, but there is a far more subtle aspect to the forces in an arch—which can be understood through the surprising consideration of a hanging chain or string. One cannot afford to have any significant tension or shear in a stone structure (which would pull or shear it apart), so the aim is to have only compression. On the left, the diagram shows three dashed lines representing the shape of a single chain or string hung from three sets of suspension points. The shape of each curve is a catenary, from the Latin *catena*, for a chain. Assuming these strings are fully flexible, there will be only tensile forces (T) in the strings. If now, we invert these curves to the solid lines above, *there will be only compression in the arches* (C), assuming that these now represent solid structures and, like the tensile stress in the chains, this will act exactly along the catenary shaped arch. Thus, catenary shapes of stone arches will contain no or minimal tension or shear and thus be the most stable geometry.

Figure 5.33 Robert Hooke and the hanging chain shape: the catenary. (Rita Greer, *Robert Hooke, Engineer*, 2009. Free Art License 1.3, Wikimedia)

It was Robert Hooke (1655–1705), Figure 5.33, scientist, Renaissance scholar, and contemporary of Isaac Newton, who discovered this catenary arch principle. He was somewhat nervous about announcing his discoveries and so published them first in anagram form; later on, if they turned out to be valid, he could then draw attention to the principle stated in the earlier published, and hence dated, anagram. His most well-known law, Hooke's Law, states that the force required to extend a wire or spring is proportional to the amount of extension, i.e., as the tension, so the extension, or "*Ut tension, sic vis*" in Latin. He published this first as the anagram *ceiiinosssttuv*. Similarly, for the catenary arch idea in Figure 5.32, he first published the anagram *abcccddeeeeefggiiiiiiiillmmmmnnnnnooprrssstttttttuuuuuuuux* before revealing it to be "*Ut pendet continuum flexile, sic stabit contiguum rigidum inversum*", which translates loosely to, "A continuous hanging flexible, is upsidedown similar to a rigid stand."

This principle has far reaching implications, not only in enabling arches to be constructed in safer geometries, but by adding weights to the chain enables the safe design of more complex stone structures. Note also that the chain supports in Figure 5.32 can be even further apart, so that the same principle applies to low arches—which we illustrate later in this chapter with reference to three stone bridges across the river Thames.

5.3.2 Buttresses

The extended conclusion from the catenary discussion in the last section is that the inverted catenary line of force should be contained not only within the arch itself but also within the supporting stonework. In Figure 5.34, we show one half of an arch that is not necessarily in the catenary geometry. However, so long as the line of thrust LT lies within the stonework, and preferably within the central third of the arch, i.e., between the dashed lines, the arch will be stable. The important point now is that, at the base of the arch, the support will have to resist the thrust (T) created by the arch, which can be resolved as the two components, W and HT, i.e., the weight of the half-arch (W) and the horizontal thrust (HT). (Note that in this explanation and in Figures 5.34–5.36, the symbol T is used to represent the compressive

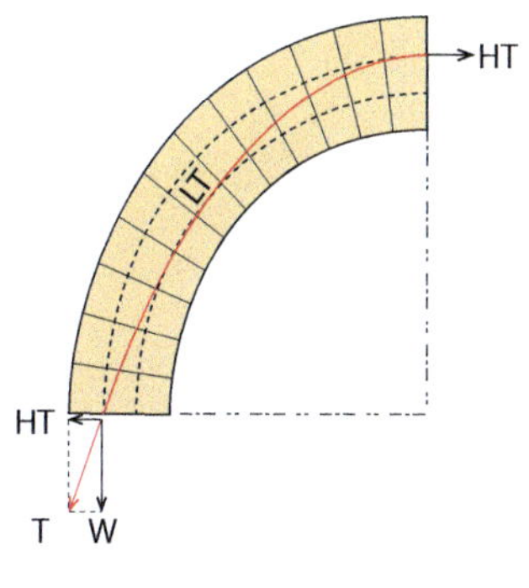

Figure 5.34 The line of thrust, LT, must lie within the stone arch for it to be stable, ideally within the central one third of the arch. Also, the thrust, T (comprised of the weight, W, and the horizontal outward thrust, HT) must be resisted, i.e., within the wall and other stonework supporting the arch.

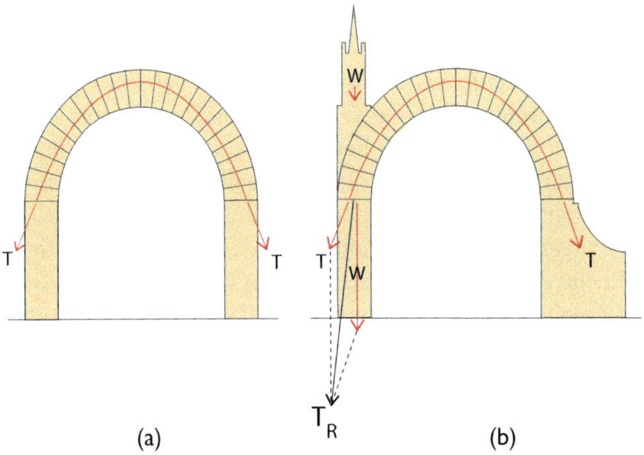

Figure 5.35 (a) The thrust line, T, within a complete stone arch. This arch would be stable if it were not raised on the pillars because the line of thrust (T) would then lie completely within the arch. But, because the arch is raised on pillars (i.e., it is stilted), the line of thrust now lies outside the stonework and the structure will collapse if this thrust (T) is not resisted. In (b), the thrust (T) is counteracted by two different methods. On the left side, an additional weight, W, is provided by the added tower which is built on the left side of the arch, so the resultant line of thrust, now T_R, *remains within the arch support*. An alternative stability option, shown on the right-hand side of the arch, is to add extended base stonework for the same purpose: i.e., to keep the line of force, T, within the stone structure.

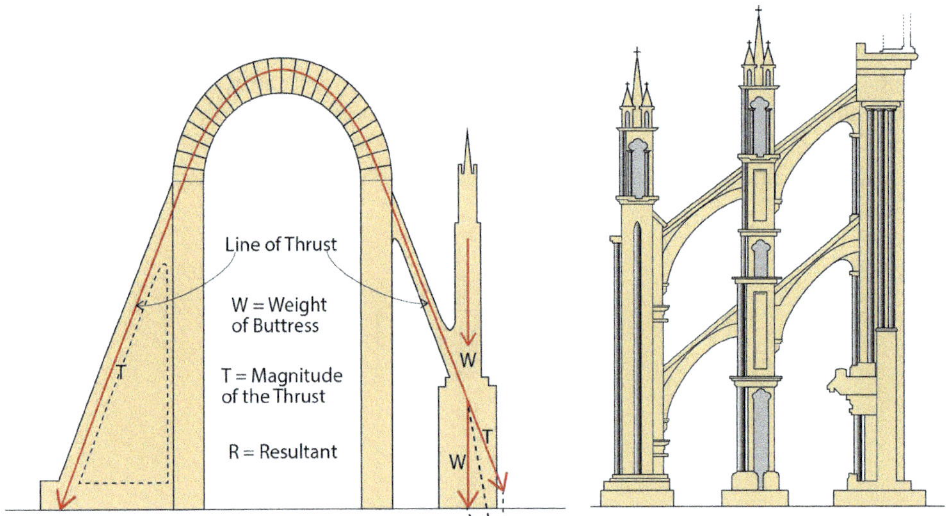

Figure 5.36 (a) A stone arch supported on high piers. On the left side, stability is ensured by the use of a large buttress, much of which is not required to resist the thrust line, T. On the right side, the thrust is channelled down only the necessary part of the buttress—leading to support members known as flying buttresses. (b) Use of this principle illustrated by the flying buttresses at Reims Cathedral, France.

thrust and not the tension discussed in relation to Fig. 5.32.) To avoid collapse of the arch, some form of stonework is required that can resist the outward thrust, as is illustrated in Figure 5.35(b).

In Figure 5.35(a), a stone arch built on pillars, known as piers, is represented. Note now that the line of thrust would exit the arch above ground level—which would make the arch unstable because there is no resistance to the thrust. There are two stabilising options shown in Figure 5.35(b). A widely used architectural solution is indicated on the left-hand side of Figure 5.35(b): a small tower of weight, W, is built above the arch so that the resultant force, T_R, remains within the stone structure to ground level. The alternative solution, shown on the right-hand side of Figure 5.35(b), extends the stonework at ground level to form a buttress enabling the resultant force, T, to reach the ground within the stonework. However, this solution is architecturally clumsy, not to mention the excessive use of stone.

In Figure 5.36(a), the inelegant solution, using a large, solid block of stonework that is contiguous with the arch, is shown on the left side. However, because the thrust, T, is directed specifically along the thrust line, much of the buttress is unnecessary and the dashed portion could be removed without threatening the integrity of the structure. On the right of the arch, this concept has been used and the structure has been reduced to a thin span of stone which transmits the thrust of the arch to a free standing buttress known as a flying buttress. In this example, the use of an additional weight to deflect the thrust line downwards, as shown on the left side of the arch in Figure 5.35(b), has been used to further reduce the footprint of the buttress. Note that the semi-circular arch is thick enough to contain the catenary form of the thrust line within its central third, see Figure 5.34.

Examples where these mechanical principles have been applied are illustrated in Figure 5.37(a) and (b). In the Lincoln Cathedral case, the flying buttresses lead the thrust line to the heavy ground buttresses; whereas, in the Salisbury Cathedral case, the flying buttresses from the clerestory level lead to stonework within the Cathedral at a lower level. The mini-arches at ground level in the external buttresses of Hereford Cathedral, whilst being delightfully idiosyncratic, do follow the mechanical principle that we have outlined in that the line of thrust will be towards the exterior of the buttress. We wonder what was in the mind of the stonework designer; was it to provide extra architectural decoration or was it to save on building stones?

Ground buttresses, Lincoln Cathedral

Flying buttresses, Salisbury Cathedral

Figure 5.37(a) Cathedral buttresses designed according to the principle illustrated in Figure 5.36.

Figure 5.37(b) Use of external buttresses to resist the outward roof forces. Note also the fleches at the top of the buttresses, which provide extra vertical forces in addition to their decorative role. Also, the small idiosyncratic half-arches near the base of the buttresses recognise the outward trend of the force vector. Hereford Cathedral, England.

So, the exterior architectural structure of churches is, firstly and predominantly, determined by structural stability considerations and, secondly, by decorative aspects—hopefully with a harmonious combination of the two. This is manifested by the exterior appearance of many churches which are dominated by the buttresses built to stabilise the internal stone arches and exterior wall, as shown in Figure 5.38 for St. Mary's Church in Hitchin, Hertfordshire. To the eye, these supports appear to be structurally in keeping with the principles outlined earlier and the scale of the building; however, the buttresses in Figure 5.39 supporting the tower of the same church seem to be out of scale with the size of the tower. Why are these buttresses so large? They were built in the 14th century before the understanding of forces and stresses was developed and so must have been stimulated by some evident instability problem, and indeed

Figure 5.38 The southern exterior of St. Mary's Church, Hitchin, Hertfordshire, dominated by the line of buttresses supporting the structure.

Figure 5.39 Substantial buttresses supporting the tower of St. Mary's Church, Hitchin, Hertfordshire.

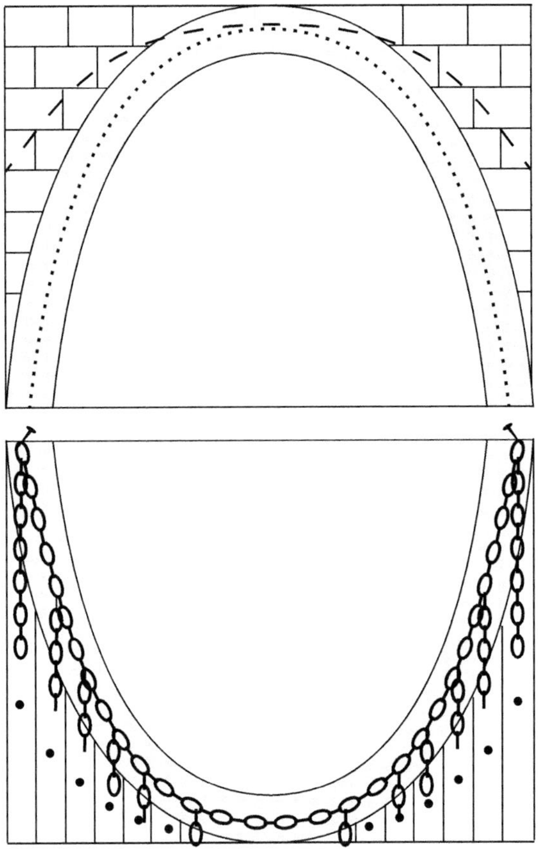

Figure 5.40 Use of the catenary principle to determine the ideal arch shape when the arch has also to support surrounding masonry. Given an initial arch shape (the dashed arc in the top diagram and the dots in the lower diagram), weights representing the extra masonry weight are added to the chain model, resulting in the modified arch (shaded in both diagrams).

that was the case: the tower was built in the 13th century but in the early 14th century was found to be moving away from the rest of the church, so the substantial buttresses were added. An interesting feature of this tower is that flints were included in the tower's building stones because it was thought that towers with flint in their stonework would not be hit by lightning.

In Figure 5.32, we showed that the shape of a hanging chain (a catenary) can be changed by widening or narrowing the two hanging points; in other words, there is a wide variety of catenary geometries for a given length of chain. This means that there is always a suitable catenary shape for a certain arch circumference between two supports. Moreover, the catenary chain shape can be modified to account for additional loading of an arch (Fig. 5.40)—either symmetrically, as in the Figure 5.40 wall loading, or asymmetrically, by adding chains to represent the load of the masonry above the arch.

* * * * *

Figure 5.41 Abbey Gateway in Chester. This 14th century gateway was the main entrance to the Abbey of St. Werburgh, which is now Chester Cathedral.

The preceding text has provided a conceptual understanding of stone arch mechanics—i.e., why a stone arch can have different shapes as a function of its height, why low stone arches can be stable, how stone arches can be stabilised by means of buttresses, and why stone arches can remain standing—even for thousands of years. Via this understanding, the west face of the sandstone Abbey Gateway in Chester, shown in Figure 5.41, can be considered with its main arch having masonry and other loading as in Figure 5.40, with the smaller open pointed arch, and the adjacent building providing the required buttressing.

* * * * *

Through the centuries there has been an inevitable mechanical evolution in arches, from the trilithon type consisting of two uprights and a cross-beam (see the Stonehenge example in the Frontispiece), to the Roman semi-circular arch, to the Gothic or pointed arch, Figure 5.42. In this case of colonnaded arches, the necessary buttressing discussed earlier is automatically provided by the adjacent arches.

We continue now with a Section on one of the main uses of the arch, i.e., bridges, with the previously described inverted catenary concept showing why bridges can have such wide, low and stable arches, quite contrary to one's initial intuition.

5.3.3 Bridges

Humanity's early settlements tended to be based on riverbanks because of the need for water, a concomitant requirement being to be able to cross the river—and so bridges began to evolve, from simple wood constructions to major stone edifices and hence the need for arches. Figure 5.43 shows a charming example of a vernacular bridge in Yorkshire. It is intriguing to imagine how the forces are distributed around the arch in this bridge: on the one hand and without the arch stability knowledge explained in the previous sections, it is easy to imagine the keystone blocks at the top of the arch falling out and the whole bridge collapsing; on the other hand and with the catenary type arch stability concept, one can have confidence in the bridge stability.

Figure 5.44 shows the Pont d'Avignon bridge in France with Gothic-type arches; Figure 5.45(a) illustrates, via an early sketch, the use of pointed arches in England at Durham; while Figure 5.45(b) shows the rounded arches of Prebends Bridge, also over the river Wear in Durham.

A subject that we will address in Chapters 7 and 8 is the weathering of stone and consequential deterioration of stone structures. Note the significant weathering of the local Carboniferous sandstone used for the Prebends Bridge across the river Wear in Durham,

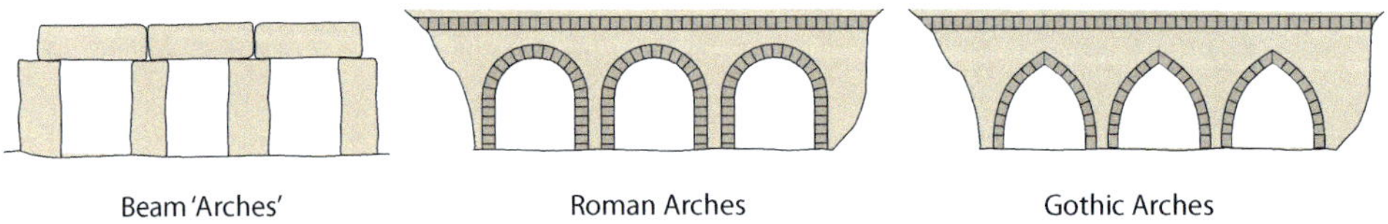

Figure 5.42 The evolution of stone arch geometry.

Figure 5.43 Bridge over Whitsundale Beck near Ravenseat in Whitsun Dale, Yorkshire National Park, England. Note that the stones directly over the stream are arranged radially, i.e., in Voussoir arch style.

Figure 5.44 The remaining portion of the Pont d'Avignon bridge (Pont Saint-Bénézet bridge) in France across the Rhône and built in the 14th century. The bridge is now a World Heritage Site. The arches are of the pointed Gothic type.

Figure 5.45(a) Old sketch of Elvet Bridge over the river Wear in Durham, showing pointed arches and that houses used to be built alongside the bridge thoroughfare.

Figure 5.45(b) Prebends Bridge with semi-circular type arches, built in 1778, over the river Wear, Durham, England. See Chapter 7, Figure 7.25.

Figure 5.46 Weathering of the sandstone used to construct the Prebends Bridge over the river Wear in Durham, England.

Figures 5.45(b) and 5.46. This is one of the three stone bridges in the centre of Durham, see the Figure 7.25 map in Chapter 7; it was built in 1772–8, forms part of the estate of Durham Cathedral, and is now a listed building. Such weathering has also severely affected the sandstone in the nearby Durham Cathedral and castle, a subject which we discuss in detail in Chapter 7. Local sandstone has also been used for the railway viaduct in Durham, England, Figure 5.47.

Figure 5.47 Sandstone railway viaduct with semi-circular arches in Durham, England.

The stone bridge over the river Ouse at St. Ives in Cambridgeshire, England, shown in Figure 5.48, was built in the 1420s and has both rounded Roman-type and pointed Gothic-type arches. Following the evolution of architectural geometry, as we describe in Section 6.9, we would expect the rounded arch portion of the bridge to be older than the pointed arch portion, but in fact it is the other way around: the rounded arches are the most recent. This is because the original arches on that side of the bridge were dismantled by Oliver Cromwell's Roundheads in 1645 in case of attack by the Royalists during the Civil War. They were rebuilt in Roman arch style in 1716. Note that the purpose of the building on the bridge was for taking tolls from travellers but it was also used for church services.

Stone bridges across the river Thames in London

> *"Sweet Thames! run softly till I end my song."*
>
> Edmund Spenser (1552–99)

The stability of the previous bridges we have illustrated, semi-circular and pointed, are intuitively stable: in other words, we do not feel that they are about to collapse. However, this is not the case for bridges with lower, more elliptically shaped arches. In order to have an

Figure 5.48 Bridge over the river Ouse at St. Ives in Cambridgeshire, England.

intuitive feeling for the stability of such bridges, we now highlight the three stone bridges that span the tidal portion of the river Thames. Bridges across this river have played a great part in the history of London and in John Stow's 1618 *Survay of London* he comments on the first stone bridge in London as follows:

> This worke, to wit, the Arches, Chappell & Stone bridge ouer the riuer of Thames at London, hauing bin 33. yeeres in building, was in the yeere 1209. finished by the worthy Merchants of London, Serle, Mercer, William Almaine, and Benedict Botewrite, principall Masters of that worke; for Peter of Cole-Church deceased foure yeeres before, and was buried in the Chappell on the bridge, in the yeere 1205.
>
> *This work, to wit, the arches, chapel and stone bridge over the river of Thames at London, having been 33 years in building, was in the year 1209 finished by the worthy merchants of London, Serle Mercer, William Almaine, and Benedict Botewrite, principal masters of that work, for Peter of Colechurch deceased four years before, and was buried in the Chapel on the bridge, in the year 1205.*

Needless to say, this original bridge no longer exists, but travelling up the river Thames from the most eastwardly bridge to the most westerly tideway bridge, a distance of 99 miles (160 km), there are 23 main bridges, starting with Tower Bridge and finishing with Richmond Bridge. However, most of these are not structural stone bridges, Table 5.1, not even the most well-known one, Tower Bridge, although it is clad in Portland stone and Cornish granite. The locations of the three stone bridges are indicated in Figure 5.49 and we pay tribute to the many engineers involved in the construction of these bridges that they still survive—given the difficulty of ensuring good foundations in the river sediment and soil beneath, the continuing river scour, plus constant traffic vibrations.

In the following descriptions (and travelling upstream along the river Thames), we provide illustrations of the stone bridges, namely Putney Bridge, Kew Bridge and Richmond Bridge. When studying the photographs of these bridges, try to imagine the inverted catenary shapes (see Fig. 5.32) of the individual arches that enable them to be stable.

Table 5.1 Main Thames bridges from east (nearest the sea) to west (furthest extent of the tide), excluding railway bridges and footbridges (the rows relating to stone bridge are shaded)

Bridge name	*Construction date of current bridge*	*Construction material*
Tower Bridge	1886–94	Metal with Portland stone and granite cladding
London Bridge	1967–72	Metal and concrete
Southwark Bridge	1912–21	Metal, granite piers
Blackfriars Bridge	1860–9	Metal, first with elliptical style arches
Waterloo Bridge	1937–42	Concrete clad in Portland stone
Westminster Bridge	1854–62	Metal, the oldest bridge in central London, Coade stone lion on south side (see Section 3.9.5)
Lambeth Bridge	1929–52	Metal, granite-faced piers
Vauxhall Bridge	1895–1906	Metal, granite piers
Chelsea Bridge	1934–37	Metal, granite-faced piers
Albert Bridge	1871–3	Metal, iron and concrete piers
Battersea Bridge	1771–2	Metal, granite piers
Wandsworth Bridge	1936–40	Metal, granite-faced concrete piers
Putney Bridge	1973–6	Aberdeen and Cornish granite, concrete piers, five spans, church at each end
Hammersmith Bridge	1884	Metal, concrete piers clad in Portland stone
Chiswick Bridge	1933	Reinforced concrete clad in Portland stone, concrete piers
Kew Bridge	1903	Stone, three elliptical type arches, granite from Cornwall and Aberdeen
Richmond Lock	1894	Metal, concrete piers clad in granite
Twickenham Bridge	1933	Stone clad concrete, piers on footing of compressed cork
Richmond Bridge	1774–7	Stone, Portland limestone, five elliptical type arches, oldest surviving Thames tideway bridge

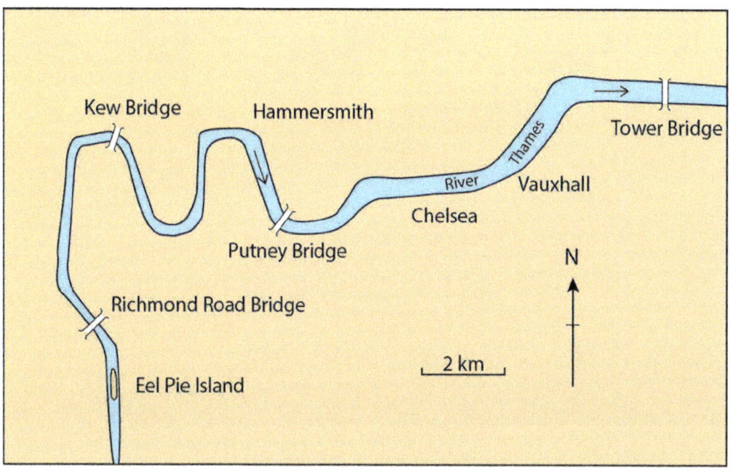

Figure 5.49 The river Thames, Tower Bridge and the three stone bridges highlighted in this section.

Putney Bridge

Derivation of Putney: From the Anglo-Saxon chief Putta, Putney meaning Putta's landing.

It is said (Roberts, 2005) that the first Putney Bridge would never have been built if the journey of Prime Minister Sir Robert Walpole to London, which required a ferry across the river Thames at Putney, had not been interrupted because the local ferrymen refused to leave the Swan pub. Walpole subsequently petitioned Parliament for a crossing in the 1720s and in 1726 an Act of Parliament authorised the construction of a bridge. This timber bridge was, however, superseded by the current stone bridge in 1884.

As can be seen in Figure 5.50, the current Putney Bridge is an elegant bridge with five elliptical style arches. It was constructed with granite (500,000 $ft^3 \cong 8495\ m^3$) from Aberdeen and the Prince of Wales' own quarries in Cornwall, Figure 5.51. The bridge was designed and its erection supervised (starting in 1882) by Sir Joseph Bazalgette, well remembered for his metropolitan drainage scheme in London. The contractual cost of construction was £240,433 and the bridge was officially opened by the Prince of Wales in 1886. The Prince laid a memorial stone of Cornish Cheesewring granite, which weighed six tons (5.4 tonnes) and was bedded at a level of 18 inches (0.46 m) above the Trinity highwater mark. The stone is inscribed "This stone was laid by their Royal Highnesses the Prince and

Figure 5.50 The elegant low arches of Putney Bridge over the river Thames at low tide.

Figure 5.51 A xenolith in the Putney Bridge granite. As granite magma intrudes upwards within the Earth's crust, pieces of the surrounding rock inevitably become incorporated within it. These fragments are termed xenoliths (i.e., rock strangers) and, depending on their composition and the time they remain in the magma before it cools and solidifies, they may melt and become part of the magma—or may survive, as shown in this photograph.

Princess of Wales, 12 July 1884." The structure of the bridge seems to have survived well over the succeeding years although, in order to account for increasing traffic, it has been widened twice, most recently in 1933. However, such widening does not affect the stability of the arches. Note that Putney Bridge is close to the start of the annual Oxford and Cambridge Thames boat race.

Putney Bridge has the distinction of being the only bridge in Britain to have a stone church at each end (All Saints' Catholic Church, Fulham, on the north side, and St. Mary's Church, Putney, on the south side), Figure 5.52, the significance in this book's context being that the towers of both of these parish churches were built using the easily recognisable Kentish ragstone (a limestone)—see Section 3.4.5 for a description.

These churches have experienced considerable structural changes since the 14th century but the Kentish ragstone towers seem to have survived the vagaries of the English climate, although the stone does eventually break down, Figure 5.53(a). The ragstone contains fossils and a particularly impressive ammonite of Cretaceous age in St. Mary's Church wall is shown in Figure 5.53(b).

Kew Bridge

The current Kew Bridge with its three elliptical style arches, Figure 5.54, was opened in 1903 and is the third bridge at this site and, like Putney Bridge, it is also built with granite from Cornwall and Aberdeen. Some close-ups of the granite characteristics are shown in Figure 5.55.

All Saints' Church, north end of Putney Bridge

St. Mary's Church, south end of Putney Bridge

Figure 5.52 The churches at each end of Putney Bridge. In both cases the towers were constructed using Kentish ragstone.

(a) Weathering of the Kentish ragstone, All Saints' Church

(b) Fossil impression of an ammonite in the Kentish ragstone, St. Mary's Church

Figure 5.53 Illustrations of Kentish ragstone features. (a) Weathering along weaknesses in the stone. (b) Cretaceous ammonite impression built into the wall of St. Mary's Church, south end of Putney Bridge.

The surfaces of the granite at the sides of the bridge approaches have a 'rough-hewn' texture, as can be seen in Figure 5.55. This is because the granite blocks have not been smoothed like the Putney Bridge granite, see Figure 5.51: in fact, three examples of the split 'plug and feather' holes used for extracting the Kew Bridge granite can be seen in

Figure 5.54 Kew Bridge over the river Thames.

(a) 'Rough hewn' granite texture

(b) Plug and feather holes

(c) Pegmatitic vein (orange)

(d) Xenolithic inclusion

Figure 5.55 Some characteristics of the Kew Bridge granite. (a) Texture of the 'rough hewn' stone. (b) The three small vertical grooves at the lower edge of the upper block are the remains of the 'plug and feather' hand worked method of splitting the *in situ* granite, see Section 4.3, Figure 4.3(b). (c) The formation of the light-coloured pegmatitic vein occurs as the granitic magma cools and solidifies. Remnant magmatic fluids under high pressure can fracture the granite, intrude along the fracture, and then cool and solidify to form the pegmatitic vein consisting mainly of quartz and feldspar. (d) The explanation of xenoliths is in the caption to Figure 5.51.

Figure 5.56 Excellent example of a Voussoir-type arch at the entrance to the bridge's pedestrian tunnel at the south end of Kew bridge.

Figure 5.55(b), this 'plug and feather' technique having been noted in the caption to Figure 4.3(b) in Section 4.3.1 on quarrying methods. Other features of the granite are illustrated in Figure 5.55(c) and (d), i.e., the presence of pegmatitic veins and xenoliths, although these are not frequently observed.

The photograph in Figure 5.56 shows an excellent example of the Voussoir-type arch, the geometry of which has been illustrated in Figure 5.31. Note how the top edges of the blocks have been arranged with horizontal and vertical edges to create a pleasing geometrical arrangement.

Richmond Bridge

Richmond Bridge, Figure 5.57, is the oldest surviving Thames bridge in London and is towards the upper extent of the tideway. It is a five-arch bridge built of sedimentary stone, with the arch peripheries faced with Portland stone, Figure 5.58. Other stones have also been used, as for example at the entrance to the north-east pedestrian tunnel, Figure 5.59, where the dark, well-bedded sedimentary stone is much more susceptible to weathering and erosion resulting in the etching out of the bedding planes, which can be clearly seen to be parallel to the upper and lower surfaces of the blocks.

The bridge was built in the years 1774–7 and widened in 1937–40, plus reducing the height of the 'hump' in the middle of the bridge. Built into the original Parliamentary Act for building the bridge was a severe penalty for anyone damaging the bridge structure, i.e., transportation "to one of His Majesty's Colonies in America for the space of seven years."

Figure 5.57 The five-arched Richmond Bridge over the river Thames.

(a) (b)

Figure 5.58 (a) The central arch of Richmond Bridge. (b) A close-up of the peripheral facing of Portland stone.

Figure 5.59 Weathering of the sedimentary stone used for the construction of Richmond Bridge, as viewed at the entrance to the north-east pedestrian tunnel.

* * * * *

5.3.4 The Wellington arch: a triumphal arch

In addition to the utilitarian role of the arch for supporting windows, doors, bridges and a variety of other architectural features, we also note at this point that triumphal arches have played their part throughout the centuries. Early examples are Hadrian's and Trajan's triumphal arches from Roman times and the style has been copied in many countries. Good British examples of this style are Wellington Arch designed by Decimus Burton which is located to the south east of Hyde Park and at the west end of Constitution Hill in central London, and Marble Arch designed by John Nash which is now at the west end of Oxford Street, also in London. The majesty of Wellington Arch is shown in Figure 5.60.

The perturbed history of both arches is well written up in an English Heritage booklet titled *The Wellington Arch and the Marble Arch.* Marble Arch is clad with marble from the Italian Carrara quarry, a building stone which we have discussed elsewhere in the book (Section 4.3.3), so we concentrate here on the Wellington Arch. In brief, after the Napoleonic Wars, which took place at the beginning of the 19th century, it was decided

Figure 5.60 Wellington Arch in London, built to commemorate Britain's victory over Napoleonic France and originally intended as an entrance to Buckingham Palace. The large sculpture above the arch is known as a 'quadriga', i.e., a four-horse chariot. It carries the Angel of Peace holding the Laurels of Victory aloft.

to have a national monument, a victory arch, at the entrance to London and located at Hyde Park Corner, which is at the south-east corner of the park where Knightsbridge meets Piccadilly. In 1816, Parliament voted £300,000 (≅ £6 million in today's money) for this monument.

The final design for the arch copies the styles of the Arch of Titus in Rome and the Arch of Trajan at Benevento, but differs because of the broad lintels, see Figure 5.60. In fact, the lintels are not load supporting and are suspended from girders made of cast and wrought iron, a somewhat ironic feature given that we have explained the structural advantages of the arch itself! In addition to controversy about the location of the arch, there was also disagreement about the statue above the monument. It started as a giant bronze equestrian statue of the Duke of Wellington, but after a trial period and relocation of the Arch in 1883, the statue was removed and it wasn't until 1912 that the new 'quadriga' was erected atop the Arch, this sculpture is the largest bronze statue in Europe weighing 38 tonnes. (A quadriga is a chariot drawn by four horses and represents triumph.) The monument is constructed with Portland stone and the adjacent statue of the Duke of Wellington is mounted on a granite plinth.

5.3.5 Roof vaults

Vault: *Arched roof, set of arches centred on a point or line.*
Groin: *Edge formed by intersecting vaults.*

We now come to the use of stone in supporting roofs, particularly with reference to churches and cathedrals. The vault or roof is an important component of a stone building, whether it is the roof itself or whether it provides the support for a roof made of another material. There was a continuous evolution of stone roof arches from before the Roman times up to the Middle Ages, so much so that the vaults became extremely varied and geometrically complex. In this section, we provide a basic understanding of stone vault geometries so that readers will be able to recognise the different stone components and hence appreciate the nature of the vaulting that they see. In Figures 5.61(a) and (b), we illustrate stages in vault development from the early Roman arch (which we have already described in Section 5.3.1 as the Voussoir arch) to the more sophisticated sexpartite and tierceron vaults. These diagrams help one to understand the geometrical structure of the stone arches in the photographs that follow.

The Barrel Vault is just an extension of the Roman or Voussoir arch to create a tunnel, noting that the individual stone courses are locked together via the stone bonding. The Romans used this type of vault extensively—for their temples, theatres and public baths. When two such Barrel Vaults cross, groins are formed where the tunnels meet, leading to the intersecting groined vault, and so a building with several successive groined vaults has the pattern shown in the lower part of Figure 5.61(a).

To overcome geometrical difficulties with the basic groined vault, a single pointed arch was added between opposing pillars, this being known as a quadripartite (four parts) vault. When this is further enhanced by a central additional transverse arch with separate supports, it becomes a sexpartite (six parts) vault. When further ribs are added for structural or decorative reasons, it is a tierceron vault, see the lower example in Figure 5.61(b). The more complex arrangements led to the term 'rib vaulting' and a variety of decorative variations on the basic structural theme.

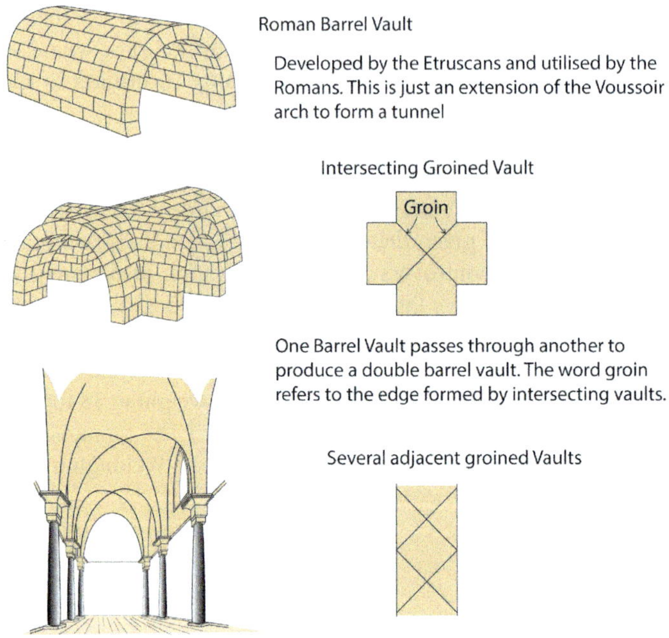

Figure 5.61(a) The geometry of barrel and groined vaults.

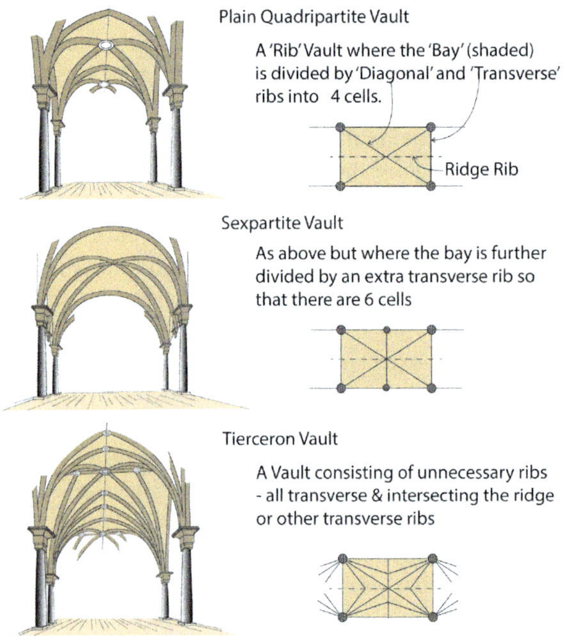

Figure 5.61(b) The arch geometry of quadripartite, sexpartite and tierceron vaults. The line drawings on the right hand side are plan views, i.e., the lines indicate the pattern of arches when viewed from above.

The picture of part of the Norman crypt in Figure 5.62 illustrates the intersection of two semi-circular arches. These two arches being built diagonally across the passageway have a larger radius and hence height than if they had been built directly across the passageway. That is why the single arch in the foreground and built from the same level has to be pointed in order to reach the same height as the diagonal arches. An alternative solution would be to make the shape of the intersecting arches more elliptical so that their intersection height would be lower. The diagrams and text in Figure 5.61(a) further illustrate these points.

We can now feel confident in interpreting the style and associated structural reason for the vaulting of St. Giles' Cathedral in Edinburgh (Fig. 5.63), Salisbury Cathedral (Fig. 5.64) and Chichester Cathedral (Fig. 5.65) with their diagonal and transverse ribs all illustrating the 'plain quadripartite' vaulting diagrammatically shown at the top of Figure 5.61(b).

Figures 5.63–5.65 are photographs of the vaults of the three cathedrals in Edinburgh, Salisbury and Chichester, having four roof spaces in each arching unit. Also, in the first figure in this book, Figure 1.1, we showed the exterior of the Jewel Tower opposite the Houses of Parliament in London, which was built originally in 1365–6, but there have been significant modifications since then. We are now in a position to consider the interior and the type of roof vault which was used for this stone construction. The roof of the ground floor, 25 × 13 ft (7.5 × 4 m), is shown in Figure 5.66, together with the roof arch plan. It can be seen that the ribbing is similar to the basic quadripartite style but extra ribs have been added making it the tierceron type.

Problems could be encountered where the masonry blocks met at the top of arches, so keystones, known as bosses, became particularly important, Figure 5.66. Making them large and heavy was an advantage because their weight counteracted the tendency of the lateral stress in the ribs to cause the tops of the arches to rise. The bosses also provided an

Figure 5.62 Norman crypt of St. Mary's Church, Warwick, built in 1123 and illustrating the use of two intersecting semi-circular arches to support the crypt roof plus the structures above, see the diagram in Figure 5.61(a). (Note that the architectural style of the window is later than that of the arches.)

Figure 5.63 St. Giles' Cathedral, Edinburgh, Scotland.

Figure 5.64 Salisbury Cathedral, England, 13th century vault.

Figure 5.65 Chichester Cathedral, England, 13th century vault.

(a)

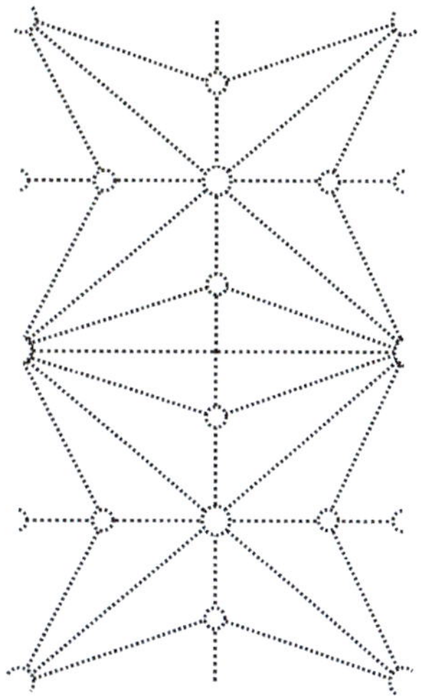

(b)

Figure 5.66 (a) The ceiling of the lower floor of the Jewel Tower, Westminster London, built in the mid-14th century. Note the carved Reigate stone bosses along the central line. (b) The tierceron type vault arch plan.

opportunity to introduce decorative features and examples from Lincoln Cathedral and St. Mary's Church, Warwick, are shown in Figures 5.67 and 5.68. Many of the bosses have survived over the years because they are, by definition, in extremely inaccessible locations and hence not so susceptible to damage, theft or destruction. Also shown in Figure 5.68 are examples of 'flying ribs' at St. Mary's Church; these were used when the requirement for higher windows and hence a flatter roof meant that the traditional type of vaulting needed to

Figure 5.67 Replica of the 'Coronation of Our Lady' roof boss from Lincoln Cathedral, *ca.* 1260 AD.

Figure 5.68 Examples of roof bosses and flying ribs at St. Mary's Church, Warwick.

Figure 5.69 Interior of King's College Chapel, Cambridge, England, built in the 16th century.

be supplemented by additional support—i.e., the individual flying ribs linking the mid-line of the roof directly to pillars in the sidewall.

As the vault styles evolved, more complex geometries developed by adding liernes, which connect any point on one rib to any point on another rib (and not necessarily providing any extra support)—so as to produce attractive geometrical patterns known as reticulation from the Latin word '*reticulum*', meaning a small net. These patterns eventually reached a zenith (figuratively and literally) in the roof of King's College Chapel in Cambridge, Figure 5.69. Although this fan vaulting appears complicated (and it is!), the structure follows the same principles that we have outlined earlier, i.e., the lines of force caused by the weight of the masonry within the vault should all lie within the

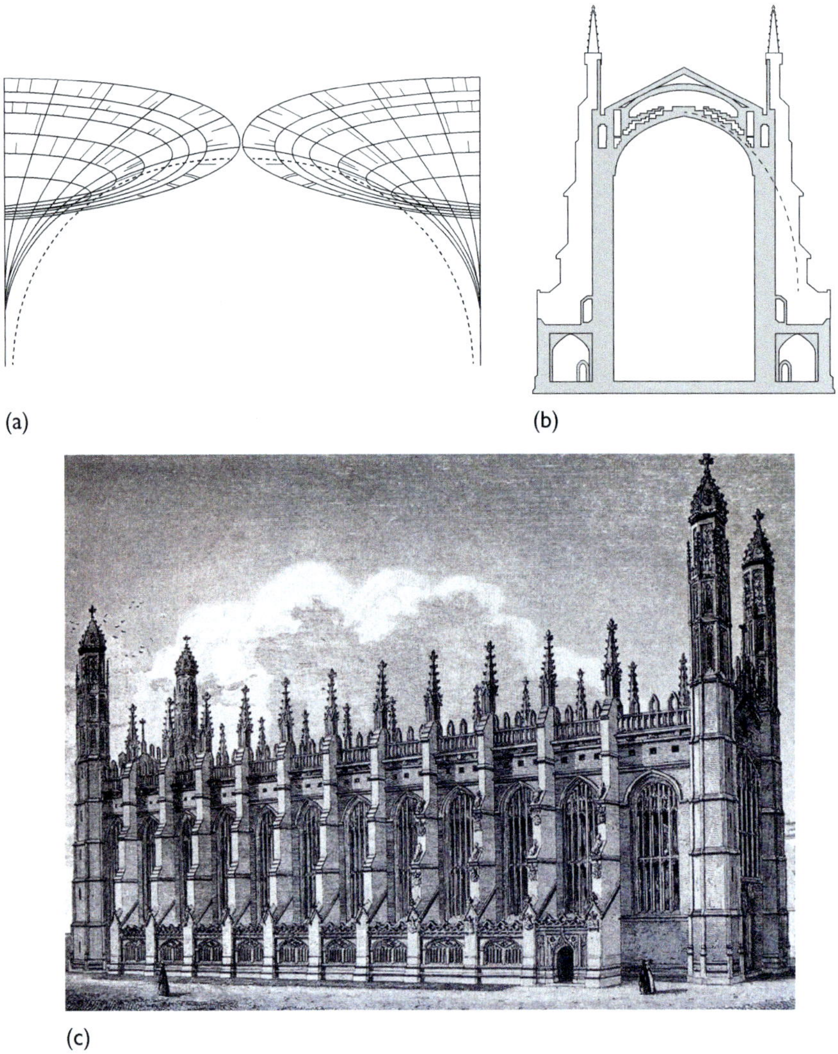

Figure 5.70 (a) Lines of thrust caused by the weight of the masonry contained within catenary curves which are then (b) and (c) sustained by the external buttresses of King's College Chapel, Cambridge. The white magnesian limestone of Permian age from Yorkshire was mainly used, intimated by the whiter faces of the right-hand buttresses, but the chapel was completed using Jurassic oolitic limestones from Northamptonshire. (Line engraving by R. W. Smart, (1826) after R. B. Harraden. Wellcome Collection. CC BY 4.0)

masonry—as determined by a catenary geometry—and then be contained within external buttresses, Figure 5.70.

* * * * *

For those readers of a more technical disposition, the book by J. Heyman (1995) titled *The Stone Skeleton: Structural Engineering of Masonry Architecture* is highly recommended for further reading on the technical aspects of masonry construction and stability. The title is astute because a skeleton can be either an endoskeleton (i.e., an internal one, as in humans) or an exoskeleton (i.e., an external one, as in crustaceans, such as a crab)—and the masonry of stone buildings is of both types, providing internal and external stability.

5.4 CASTLES AND CATHEDRALS

The subjects of castles and cathedrals are taken together in this section because they represent large scale combinations of the stone pillars, walls, arches and roof vaults described in the previous sections of this chapter. However, the roles of these two types of structure are diametrically opposed: castles are defensive structures built to keep intruders out and the inhabitants safe, whilst cathedrals are receptive structures built to attract churchgoers for worship with the architecture reflecting a religious purpose.

5.4.1 Castles

Having discussed pillars, walls and arches, we continue with castles and the Robert Frost theme of "Before I built a wall I'd ask to know what I was walling in or walling out." Recalling the closed polygonal shapes of the city walls illustrated previously for Roman London and Chester, Figure 5.26(a), with castles we now 'tighten the wall loop', firstly, with the external castle wall and, secondly, with the 'keep'—i.e., the building of last retreat within the castle walls. In Britain, stone castles began to appear in the 12th century with this geometry consisting of a square stone keep surrounded by a curtain wall having an external moat, the whole architecture being defensive in nature. Natural topography could be of significant help when locating the castle—such as on the banks of rivers, e.g., Chepstow Castle, Figure 5.71, or on elevated ground, e.g., Edinburgh Castle and Stirling Castle.

The castle layouts varied and evolved as the methods of attack increased in severity, e.g., the parapet geometries were optimised with merlons and crenels (the small pillars and

Figure 5.71 Chepstow Castle, located on the river Wye in Wales.

gaps along the battlements) and the battlements themselves were 'machiolated' so the parapet extended out from the wall, enabling the defenders to be able to be directly above the attackers. Also, the rectangular and circular castle walls evolved to more complex shapes so that the defenders could fire more directly at the attackers.

There are many old castles in Britain and they are all potentially of interest because they are made of stone. There were around 600 castles in Wales, and more than 100 of these are still standing. In fact, the oldest surviving post-Roman stone fortification in Britain is Chepstow Castle (Castell Cas-gwent) on the river Wye at Chepstow, Monmouthshire, Wales, and is shown in Figure 5.71. Located on a spur of rock above cliffs on the river Wye, it was built in the 11th and 12th centuries to guard the English border, was expanded in the 12th and 13th centuries, and declined in importance during the 13th century and afterwards, until the English Civil War when it was besieged in 1645 and in 1648. Now, the ruins are a tourist attraction with Marten's Tower to the left and the current gatehouse on the right in Figure 5.71.

The effectiveness of the trebuchet (a device for hurling stones) in the 12th century and the English Civil War castle cannon attacks in the 17th century meant that castles were no longer impregnable and their construction effectively ceased. In fact, in 1649 Parliament ordered that some castles should be destroyed. Thus, the architectural evolution led to essentially impregnable castles—until the advent of heavy cannon power which threatened the very integrity of the castle walls, as indeed did underground mining below the walls. The use of earthen walls against the outside of the castle walls was developed as a protection against the vibrations set up by cannon attack but this has dubious value, although, as a form of earthquake defence, the walls of some Japanese castles do curve outwards at the base. Note that besieging a castle took time, especially if the intention was to starve out the occupants, so more direct methods were necessary, such as mining beneath the walls or, indeed, direct attack.

Following the previous introductory text on castles, let us now examine their development in more detail. The first major castles were built after the Norman Conquest of Britain in 1066 AD by William I, who ensured that castles dominated every Saxon town. In fact more than 100 castles had been built between 1066 and 1086 AD, the date of the Domesday survey. The Norman knights who built and lived in the castles had a duty to the king who, in turn, had a duty to help them. However, the first two centuries of Norman rule were turbulent times and so many castles were built originally in the 11th century, Figure 5.72, the nature of each castle being a function of geography, security and available building resources. The *Castellarium Anglicanum*, published in 1983, lists over 1500 castle sites in England and Wales. The remains of all these castles vary: some are still in reasonable condition, some have suffered significant deterioration, whilst others have simply disappeared.

The basic motivating principle of protecting the inhabitants from external attack for potentially long periods of time automatically leads to certain key principles, such as the need to be able to resist attack through strong defensive walls and for a continuous supply of drinking water to be available, preferably a well. This then leads directly to the best site being on advantageously high ground, e.g., Lincoln Castle, to the West of Lincoln Cathedral, illustrated by the old sketch in Figure 5.73(a), ideally surrounded by water, and to the use of stone for the structure itself. For these reasons, excellent sites are on high ground and at the bend of a river—such as Durham Castle, which is next to Durham Cathedral (discussed in some detail in Section 7.3). Another good example is Edinburgh Castle, Figure 5.73(b), built on the dolerite plug of an extinct Carboniferous volcano, having steep sides to the south, west and north. To obtain water, a 92-ft (28 m) deep well was excavated, although this was not always a reliable source. Stirling Castle, also in Scotland, is similarly located on a rock outcrop, Castle Rock, an igneous intrusion which has been scoured during a succession of ice ages into a formidable castle location with cliffs on three sides.

Figure 5.72 Locations of major castles in England at the end of the 12th century.

Figure 5.73(a) Old sketch of the elevated site of Lincoln Cathedral in England.

Figure 5.73(b) Edinburgh Castle, built on the dolerite plug of an extinct Carboniferous volcano.

Where such natural high ground was not available, a motte (hill) was built, especially for the keep, see Figure 5.74. This motte in Cardiff was originally built by the Normans *ca.* 1081 AD with wooden defences and with a stone keep in the 1130s (the original keep being larger than the one in the photograph). The ground surrounding a castle's motte is known as the bailey and this in turn was surrounded by curtain walls, originally of timber and later in stone. As the years went by, the keep, i.e., where things are kept, and as the site of last retreat, evolved into a major building with several floors.

A castle is a structure built to resist attack and hence to protect its occupants. This objective was achieved by choosing defensive sites on high ground and/or with natural or artificial water barriers (e.g., Caerphilly Castle in Wales, Fig. 5.75) using thick stone (and hence fireproof) walls, protecting the entrances, and designing the geometry of the castle walls to be able to resist attack, both by having deep foundations to deter tunneling and high walls having the ability to resist attack at any location and from any direction. Ideally, there should be a succession of defences from the moat, to the surrounding walls, to the inner walls, to the keep.

The keep of a castle, the last line of retreat, was particularly well built, resulting in the keep often being the best preserved part of a ruined castle, as is the case for Rochester Castle on the river Medway in south-east England. This stone castle was built for William II (William Rufus) by Gundulf, Bishop of Rochester, with Kentish ragstone around 1087 and the keep was added later in 1127. This keep is the tallest surviving keep of its type in Europe, Figure 3.37. After a long history, including being used as resistance during three major sieges, it was finally purchased in 1884 for £6,572 by the city of Rochester and is still in remarkably good structural condition today, although the original timber floors and other

Figure 5.74 The keep at Cardiff Castle, built on an artificial mound, a motte.

Figure 5.75 Caerphilly Castle, surrounded by an artificial moat and lakes, Wales.

furnishings have long since disappeared. (We highlighted this keep earlier in Section 3.4.5 (Figs. 3.37–3.39) when discussing Kentish ragstone.) A typical example of a keep would have a storeroom and well on the ground floor, then a guardroom and chapel on the next floor. Above that would be the Great Hall, and further up the lord's private quarters and the dormitories, with the battlements above. Access to these floors was via spiral staircases in the corners of the keep.

A further example of a keep is the Tower of London with its White Tower on the north bank of the Thames (Fig. 5.76) which was built during the 11th–14th century, begun by William I to protect and control the city of London. It may have been designed by Gundulf, Bishop of Rochester—there are similarities with the keep at Rochester Castle, not least the Kentish ragstone used for their construction. Given the White Tower's age, its role as both fortress and palace, and its location on the Thames river in London, it has hosted many historical events, such as the condemnation of Richard II as a tyrant. The site is currently the home of the British Crown Jewels (but now in the Waterloo Barracks). It was Henry III in the 13th century who instructed the external stone to be whitewashed, leading to its name as the White Tower. Over the years, many changes have occurred to the stonework, including replacement (a difficult task) of some of the original Caen stone from William I's hometown in France. The Chapel of St. John in the White Tower is a superb example of Norman architecture with its sturdy pillars and semi-circular arches.

Figure 5.76 The White Tower of London viewed from across the Thames river in London.

For attacking a castle, we are familiar with films that show fights between those on ladders against the exterior walls and the castle defenders. Siege towers were another method of scaling the walls as these were high enough to overlook the castle walls and had a drawbridge system to enable access over the walls. An alternative method was the battering ram made from a tree trunk; this was particularly effective once the castle wall's outer layer of stones had been penetrated, because the rubble core did not provide much further resistance. Mining under the walls was another method: once a tunnel had been made to a sufficient distance under the castle wall, the wooden props used to maintain the stability of the tunnel were set on fire leading to the collapse of part of the wall. Note that Dover Castle in the south of England was potentially susceptible to this type of attack because it is built on easily mineable Chalk strata.

Catapults were also part of the attackers' armoury and could fire stones or flaming missiles. What had not been anticipated during the main castle building periods in Britain was the invention of gunpowder, the cannon of the 15th century onwards, and particularly the cannon of the Civil War in the 17th century. On the other hand, the simplest siege method was just to surround the castle and wait until the defenders ran out of food and/or water.[3] Thus, effective defence of a castle meant of course anticipating the various forms of attack when designing and building the castle.

There was a natural evolution in the geometry of castles as the years passed. The early Norman castles were made from wood and hence were easily destroyed, so solid stone, or at least thick rubble walls with stone facings, were necessary. Another example is the introduction of the *machicoulis*. This is the stone version of the wooden platform that was constructed to overhang the outside of the castle walls so that stones and other missiles could be dropped directly onto castle attackers. In addition to its use in earlier days, the *machicoulis* became an attractive architectural feature which was used in later castle building. Similarly, the gatehouse was extended over the years with a second portcullis and accommodation rooms. However, it is said in design work that, given the need for an object to serve a particular purpose, different inventors will end up with the same type of object—and so it was with castle evolution. By the middle of the 14th century, all the separate buildings associated with different functions were linked together into one architectural scheme, albeit with many local variations on the basic theme. Note, however, that of the approximately 1500 castles estimated to have been built in England, the majority have for all intents and purposes disappeared altogether.

The attractiveness of castle architecture extended to the bishops' castles which were actually palaces for which 'licences to crenulate' had been obtained. The crenulations are the undulations along the tops of the castle walls caused by the regularly spaced gaps, crenels, for the use of defensive weapons, see Figure 5.76. In other words, the palaces were made to look like castles even though they were most unlikely to be attacked. Note that licences to build castles in earlier times had not been easy to obtain from the Crown because they could become centres for anti-Royalists, but licences to crenulate were given during less dangerous times. For example, in the time of Edward III, 181 licences to crenulate were issued and this number reduced considerably in subsequent times. In geology, the term 'crenulation' is used to describe the microscopic folds that form when a planar rock fabric, such as a slaty cleavage, is folded by a later deformation event echoing the appearance of the crenulated battlements.

During the English Civil War (1642–51), some castles were 'slighted' by Cromwell, i.e., partially demolished so that their defences were reduced. After this, as many castles fell into ruin by being abandoned because of their unsuitability for 'domestic living' as were destroyed by warfare.

5.4.2 Cathedrals

The word Cathedral comes from the Greek and Latin cathedra, meaning chair, in particular the Bishop's seat.

Castles and churches are the best remaining evidence of British stonework from the Middle Ages, i.e., from the 11th to 15th centuries. In the last section, we noted that a large number of castles were built during this period—many of which have now disappeared, although others are still standing in relatively good condition, e.g., the White Tower of London previously discussed. The same history applies to the cathedrals and the even larger number of parish churches: many have been left in ruin whilst others have faded away leaving only archaeological traces. Because we present the development of the architecture of churches and cathedrals in Chapter 6, in this section we have concentrated on the actual building of cathedrals.

Like the castles described in the previous section, the architecture of large churches and cathedrals followed the functions and conventions of earlier buildings. The basic plan of the parish church with its west–east orientation, top of Figure 5.77, was developed further to include transepts which assisted in supporting a central spire, and the church geometry then became a partial cross, as shown in the lower part of Figure 5.77. The tower with the church bells was often moved to the west end and the chancel containing the altar at the east end was enlarged, choir seating was installed and sometimes there was a distinctive east window—all of which led to the general layout for cathedrals.

In Figure 5.77, we indicate how the size of a parish church increased during the 12th to 15th centuries. At the same time and as a result of the Norman invasion, castles (discussed in the preceding section) and cathedrals were being built. For example, and in chronological order, building of the cathedrals at St. Albans, Winchester, Ely, Gloucester, Canterbury, Chichester, Durham, Norwich, Peterborough, Rochester and Hereford was developing in the period 1066–1150. This early architecture is termed Norman/Benedictine/Romanesque: Norman because of the direct influence of the Normans following the Conquest in 1066; Benedictine because of the Benedictine monastic style; and Romanesque because the Benedictine builders who had been brought to England were accustomed to the Roman style.

The design of a large church or cathedral was developed and executed by a master mason, whose role was much greater than the modern architect because the master mason was also in charge of the building works. Masons' pay was guaranteed by law: the Statute of Labourers (1360) dictated that masons must be paid daily and sometimes their room and board was paid as well. Furthermore, there were considerable links with French expertise, e.g., William of Sens was asked to oversee the rebuilding of Canterbury Cathedral after the fire of 1174. Early churches tended to be designed mainly *ad triangulum* or *ad quadratum*, i.e., their cross-section being based either on a triangle or a rectangle, the effect being that the *ad quadratum* design resulted in greater height in relation to width. We discuss further the development of cathedral and church architecture in the next chapter, i.e., the Angevin style in the 12th century, the Gothic style in the first half of the 13th century and its further development in the second half of the 13th century and the 14th century. In the 15th century,

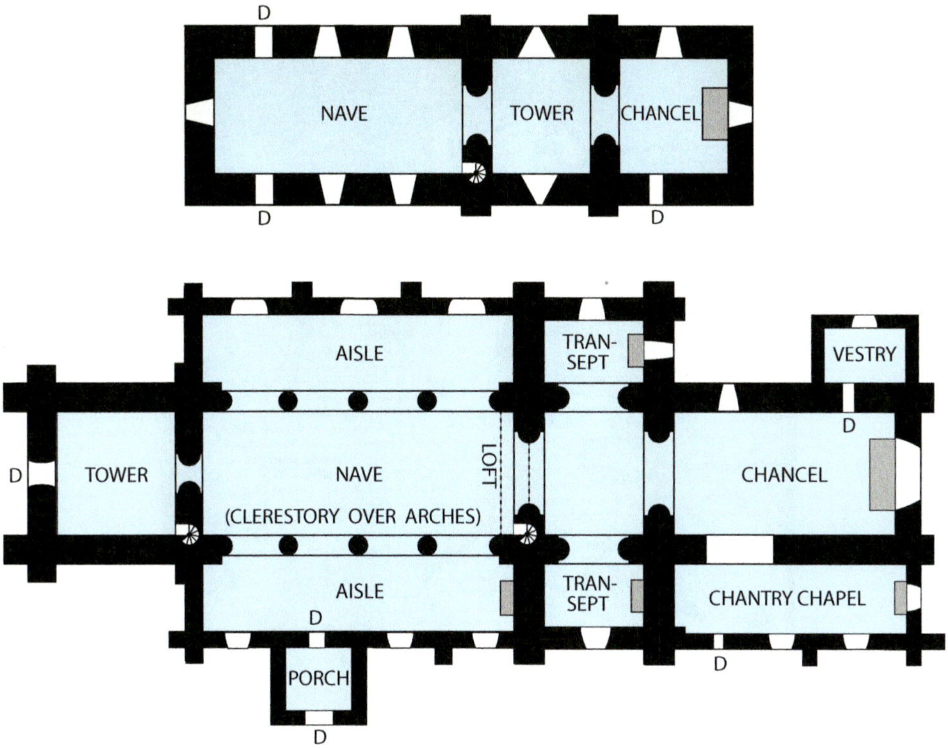

Figure 5.77 Upper Plan: parish church about 1100 AD. Lower Plan: church development about 1450 AD. (Redrawn from Brandon L. G., Hill, C. P. and Sellman, R. R. (1958) *A Survey of British History from the Earliest Times to 1939*. London: Edward Arnold and Co.)

the Perpendicular style was also taken up by the parish church builders. Early on, the decorative features of the church architecture were religiously based but later they became more naturalistically decorative, e.g., with birds and plants—see the cathedral decorative sculptures we illustrate in Section 6.6.

The long architectural evolution ended with the dissolution of the monasteries by Henry VIII in the years 1536–41. Of course, architectural development continued in the following centuries but it wasn't until the 17th century that the role of the architect was distinguished from the masons of the previous centuries—with the examples of Inigo Jones in the first half of the 17th century and Sir Christopher Wren, the architect of St. Paul's Cathedral in London (see Section 6.12.2), in the 17th and 18th centuries. We develop this architectural evolution theme further in Chapter 6 and illustrate the style of architecture in Durham Cathedral in Chapter 7.

* * * * *

It is outside the intended scope of this book to discuss the specific historical construction and masonry techniques associated with stone buildings through the ages, but those who

Figure 5.78 Building of the Tower of Babel, from a French late medieval devotional book, the *Bedford Hours*.

are interested in this subject are directed to the extensive and erudite descriptions in the 629 page book *Building in England* by L.F. Salzman, published by Oxford University Press (1952). Some indication of the 15th century masonry techniques can be gathered from the illustration in Figure 5.78 which is a, not altogether serious, contemporaneous (1414–23) representation of building the Tower of Babel.

5.5 LETTERING IN STONE

Few stoop the lettering to trace
Which time's rude hand will soon efface.
From Mary S. Cope, *Western Lands* (1852–88)

5.5.1 Introduction

In the next chapter, we discuss the evolution of architectural styles through the ages and the inevitability of the specific developments which occurred over time—because of geometrical and mechanical factors and constraints—but in the case of the evolution of lettering types, as evidenced by ancient clay and stone artefacts, the situation is different because of the many arbitrary ways in which objects, sounds and concepts can be represented by marks made in stone. The Ten Commandments that Moses brought down from Mount Sinai were written on tablets of stone, and the expression 'written in stone' is used to indicate permanence and unchangeable information. However, in the context of this chapter, it is clear that, as a result of the processes of erosion and weathering over periods of years, messages 'written in stone' can become eroded, sometimes in an unexpectedly short time, depending on the type of stone used and the information is then lost.

Writing evolved through the phases of pictograms (pictures illustrating facts), then ideograms (symbols representing ideas), and then phonetic symbols (pictures and letters representing sounds)—noting that in the latter two cases it may well not be clear what a particular symbol may mean on first encounter. An example of cuneiform writing (from the Latin word *cuneus*, meaning wedge) in stone at Persepolis in Iran is shown in Figure 5.79. Without reference to specialised knowledge, we are unable to understand the message incised into the grey Persepolis limestone.

Figure 5.79 Old Persian cuneiform writing at Persepolis, Iran.

If the interpretation of an ancient script cannot be deduced directly, then parallel inscriptions of the same message in different scripts are needed, one of which can be read. This was the case for the cuneiform script which came into existence before the end of the 4th millennium BC as *Sumerian* cuneiform. Then came *Akkadian* cuneiform in mid-3rd millennium and *Elamite* cuneiform, which was adapted from the Akkadian script. Finally came the semi-alphabetic *Old Persian* cuneiform (Fig. 5.79). However, the use of these scripts overlapped in some cases and they were eventually deciphered through the discovery of inscriptions of the same message made on a cliff face at Bisitun in Iran which were in Old Persian, Elamite and Akkadian. The British historian Henry Rawlinson (1810–95) was able to decipher the Old Persian component and then the Elamite and Akkadian scripts. Finally, bilingual scripts of the same message in Akkadian and Sumerian enabled Sumerian cuneiform, the earliest, to be deciphered by 1879. Thus, through the cuneiform style of writing which was in use for 3000 years, we can now receive messages from millennia in the past 'written in stone'.

A deciphering breakthrough of old scripts was enabled by the discovery of the granodiorite Rosetta Stone (Fig. 5.80), which was found in 1799 in Fort Julien, a ruined fort north of the town of Rosetta on the west branch of the Nile. It is inscribed with three scripts: Ancient Egyptian hieroglyphs, Demotic script and Ancient Greek—thus enabling the understanding of Egyptian hieroglyphs.

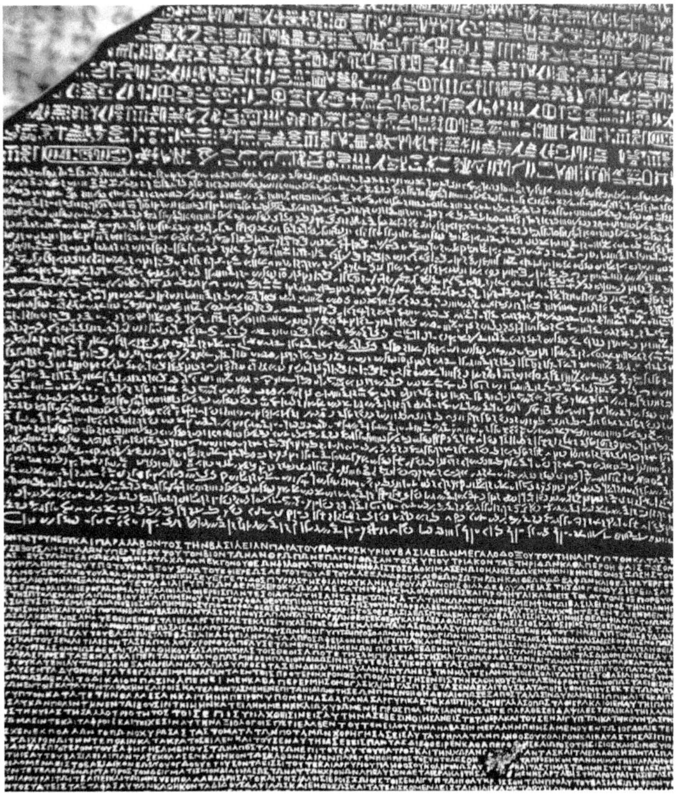

Figure 5.80 Replica of the Rosetta Stone. The original is in the British Museum, London.

The important general point is that lettering engraved in stone will generally have a long life, so that information can be transmitted through decades, centuries and millennia. But the longevity of the message depends on the life of the stone, or more precisely the life of the stone's surface. In the case of gravestones, much depends on the stone used (see Fig. 8.1) and the weather conditions: a message carved in granite will generally last much longer than a message carved in limestone. But in the fullness of time, frost, rain, sun and large temperature changes will cause erosion and hence loss of the message (as indicated in the Mary Cope quotation at the beginning of this section).

5.5.2 The Roman alphabet

The subtlety of the Roman alphabet geometry is regarded as the reason for its role as the most majestic of the alphabets. It has been said (Seaby, 1925) that the highest point of perfection of the Roman alphabet is seen in the lettering of the Trajan Column, Figure 5.81, which commemorates Roman Emperor Trajan's victory in the Dacian Wars. Although many other fonts have been developed during our current computer age (Garfield, 2010), two millennia after the Trajan Column inscription, this Roman alphabet is still used for inscriptions on the more important memorials.

A feature of this lettering is the serifs, which are the small extensions added to the ends of letter components, as a residual feature from the endings of letters when a brush was used.

Figure 5.81 Roman lettering at the base of Trajan's Column. Note that the vertical components of the letters are thicker than the horizontal ones.

Note that a font without these is known as a *sans serif* font. The serifs seem rather subtle in the Trajan Column inscription, Figure 5.81, because of the scale, but are clear in the modern charnockite (a granitic-metamorphic stone) lettering in Figure 5.82.

The sophistication of the original Roman lettering for inscriptions is also demonstrated by the 'Hadrian Inscription' (Fig. 5.83) which was erected at a height of 5 m over the Gateway of the Forum at Uriconium (Wroxeter) in Shropshire, England. The quality of the tablet writing is relatively good because it was buried for about 1600 years before being discovered in 1924 in 169 pieces. Notice that the five lines of lettering reduce in size, from top to bottom: i.e., 23, 21, 19, 17 and 15 mm letter heights. This is a 'counter perspective' feature to account for the distance each line of text would be from the reader's eye when the inscription was mounted above the Uriconium Forum Gateway. In terms of lettering style, it is

Figure 5.82 Modern incised letters with exaggerated serifs carved in charnockite (with the top half of a black pen for scale). Charnockite, in this case having a 'bush hammer' finish, is a type of highly metamorphosed granite and named after Job Charnock, the founder of Kolkata (Calcutta), whose tombstone was made of this rock type. The letters are located in the inscription at the rear of the 8 St. James's Square office building in London.

Figure 5.83 The Hadrian Inscription, 130 AD, from Wroxeter, England.

Figure 5.84 Lettering on the outside wall of St. Clement Danes Church, London. Note the enhanced serifs and the feathering texture of the lettering incisions.

considered that this inscription exhibits more imagination, freedom, rhythm and warmth than the Trajan Column inscription (Fig. 5.81), which also shows a similar gradual reduction in letter size moving from upper to lower lines of the text.

Even just the comparison of the Trajan Column inscription (Fig. 5.81) and the Hadrian Wroxeter inscription (Fig. 5.83) indicates how subtle changes in the lettering can produce different effects, especially the incised nature of the lettering, which is clear in the modern lettering in Figure 5.82. However, it is surprising that the 'Roman font' has not only survived more or less in its original form but is also currently the favoured font for the lettering on important stone plaques. A variation on the theme is illustrated by the lettering on the outside wall of St. Clement Danes Church in London (Fig. 5.84).

5.5.3 Stone lettering examples in different types of stone

In Chapter 3, we described the features of different types of building stones in relation to their geological origin; in the stone carver's trade, these features have a direct influence on the choice of stone used for different products. For example, a limestone may be easy to carve but not have a long life when exposed to the elements; conversely, a granite may be difficult to carve but have a much longer life. Visits to a graveyard will quickly illustrate the resistance of different stones to the weather through many summers and winters, granite being the most resistant. So, following the Chapter 3 sequence of granite, limestone, sandstone and metamorphic categories, we now illustrate lettering in different building stones.

5.5.3.1 *Granite*

Granite is one of the favoured stones for lettering because of its resistance to weathering and hence its longevity. The text shown in Figure 5.85 has been carved in an attractive brown charnockitic Dakota Mahogany granite plaque in St. Mary-Le-Bow Street in London, but the text is rather indistinct because of the shallow depth of the letters and the fact that the granite texture can be seen at the base of the letters. Carved letters may also be difficult to see in other types of granite, so different techniques have been used to make the letters more visible.

In Figure 5.86, the letters stand out because the surrounding stone has been removed, a technique known as relief or bas relief carving. Although the raised lettering has the same texture as the surrounding surface, the visibility of the letters is enhanced by the shadows that are created.

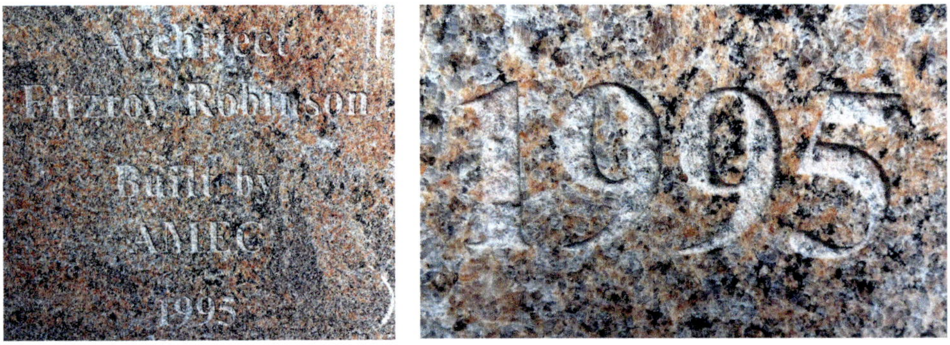

Figure 5.85 Engraved granite plaque in St. Mary-Le-Bow Street in London.

Figure 5.86 Raised lettering on a plinth in Helsinki, Finland.

Figure 5.87 The inscription in granite below the lion sculpture made of ceramic Coade stone at the east end of Westminster Bridge in London. (We featured the lion sculpture in Section 3.9 on artificial stones, Fig. 3.113.)

So, an alternative method of enhancing the visibility of the lettering is to paint or gild the letters, as in the granite block below the Coade stone lion in Figure 5.87 and the Rowland Hill plinth in Figure 5.88.

A similar technique to that illustrated in Figure 5.86 is demonstrated in the memorial to Paul Julius Baron von Reuter, shown to the left in Figure 5.89. The close up of the word

Figure 5.88 Inscription below the Rowland Hill statue in Edward Street, London EC1.

Figure 5.89 The granite monument (near the Royal Exchange in London) to Paul Julius Baron von Reuter (1816–99), the founder of the Reuters news agency.

'PRINCIPLES' to the right in Figure 5.89 was taken from the text on the reverse side of the monument.

Alternatively, the letters in granite can be deeply incised in order to provide enhanced visibility, as in the London Bridge lettering in Figure 5.90, or have a black surface, as in the Edinburgh tourist centre sign in Figure 5.91.

(a)

(b)

Figure 5.90 Deeply incised letter in the London Bridge granite inscription.

Figure 5.91 Imposing lettering above the tourist centre in Edinburgh, Scotland.

Figure 5.92 Raised lettering in a sandstone street sign in Montréal, Canada.

5.5.3.2 *Sandstone*

The grain size of sandstones is generally smaller than that of granite, so that the lettering is more distinctive in sandstones, as in the bas relief street sign carved in the Scottish sandstone imported to Montréal, Canada, shown in Figure 5.92.

In the sign above the Euston Road entrance to the British Library in London (Fig. 5.93), the visibility of the letters has been enhanced by 'texturising' the inside of the letters, whilst smoothing the surrounding stone—which is a New Red sandstone from the Permian period (298–252 million years ago) obtained from Mansfield in England.

The sans serif lettering at the street level of the 1 Poultry building in London, Figure 5.94, has been made in a block of sandstone exhibiting the 'liesegang rings' effect—which we discussed in Section 3.5 and will explain in detail in Section 7.3.4 while discussing the effect of these rings on the weathering of the sandstone structure of Durham Cathedral and its surrounding buildings.

Figure 5.93 Impressive New Red sandstone sign above the entrance to the British Library on Euston Road near King's Cross and St. Pancras international railway stations in London.

Figure 5.94 Street address of the post-modern building at Poultry and Queen Victoria Street, adjacent to the Bank Junction in the City of London financial district.

5.5.3.3 *Limestone*

Although granite is renowned for its resistance to weathering, limestone can retain the sharpness of its lettering over many years if the climate is suitable. Also, and as a general principle, not only are the grain sizes of limestones less than those of granites but they are softer and have a lighter colour so that the lettering is easier to carve and is more distinctive. Portland stone is widely used in London and an example of its 'majesty' is shown in the Cable House lettering in Figure 5.95.

The strength of black coloured Roman type lettering in Portland limestone is shown to great effect in the wall of Haberdashers' Hall in West Smithfield, London, Figure 5.96(a).

Figure 5.95 Raised Portland stone lettering in Great Percy Street, London.

(a)

(b)

(c)

Figure 5.96 Inscription in the wall of Haberdashers' Hall, 18 West Smithfield, London.

The fine nature of the deeply incised letters is clearly illustrated in Figure 5.96(b) and (c). Raised lettering in Bath stone is shown in Figure 5.97 and designs cut into Portland stone shown in Figures 5.98. The effectiveness of using a non-conventional font is demonstrated by the country trail sign in Figure 5.99.

5.5.3.4 Lettering in metamorphic stone

We have discussed metamorphic rocks in Section 3.7 and noted that this type of rock occurs as a result of the transformation of a pre-existing rock through heat and stress. The main types of metamorphic rock—slate, schist, gneiss and marble—may all be used as building

Figure 5.97 Art Nouveau lettering in Bath stone, Salisbury, England.

Figure 5.98 Coat of arms design in Portland stone on the Royal Opera House, Covent Garden, London.

Figure 5.99 Walking route sign with a modern font on Old Pale Hill near Chester, England. There are signs on the hill indicating the views of Cheshire, Derbyshire, Lancashire, Shropshire and Staffordshire in England, and Denbighshire and Flintshire in Wales.

Figure 5.100 Lettering in marble, Montréal, Canada.

stones. Figure 5.100 shows the effectiveness of marble and the associated lettering for a building in Montréal.

In the case of the majestic Finlandia Hall on the Töönlahti Bay near the centre of Helsinki, the architect Alvar Aalto chose Carrara marble from Italy, not only for the exterior of the building but also for the sign on Mannerheim Street, Figure 5.101(a). A close-up of the

(a)

(b)

Figure 5.101 (a) Lettering in Carrara marble at the Finlandia Hall in Helsinki, Finland. (b) Detail from (a).

cut-out letter N from this sign is shown in Figure 5.101(b). The orientated texture in both the Montréal and Helsinki marble examples is most attractive and the reason for the widespread use of Carrara marble as a decorative stone. Indeed, other marbles are also used to great decorative effect, e.g., Figure 5.102 showing a plaque in St. Botolph without Aldgate Church, London.

If we may for a moment consider brick as an artificial metamorphic stone, beautiful designs can be created by cutting away portions of a brick wall to create *en bas* relief lettering, Figure 5.103.

Figure 5.102 Marble plaque in St. Botolph without Aldgate Church, London.

Figure 5.103 Skilful brick design in Chester, England.

5.5.4 Notes on old and new methods of carving and stone lettering in stone

Hand lettering using a chisel requires considerable practice and skill which is achieved over many years working in the trade. But there is one guiding principle that applies to breaking rock, whether it be for stone lettering, blasting rock on the large scale or designing the cutters on a tunnel boring machine: and that is 'breaking to a free face'. In sequential rock blasting, a series of boreholes are drilled and then charged with explosives which have time-delay detonators. Once the explosive in the first borehole has been detonated, the damage has created some free, unstressed surfaces in the rock mass which then enhance the fracturing induced by the subsequent detonation in the second borehole, and so on for the remaining boreholes. This use of delay detonators in boreholes is known as a 'burn cut'. So ensuring a suitable geometry for such boreholes enhances rock blasting. Similarly, for a tunnel boring machine, which has a series of cutting devices (e.g., discs or buttons) mounted on its rotating cutting head, the same principle is used—as the head rotates, the individual, but slightly offset cutters, follow each other around so that rock breakage occurs into the groove made by the preceding cutter.

For cutting out the letters in a building stone, exactly the same principle is used: an initial groove is created along the centres of the letter lines, and then the exact form of the letters is generated by chiselling *such that stone breakage occurs towards the initial 'free face'*, i.e., the initial groove (Fig. 5.104). This not only makes the whole process easier, but it also reduces the chance of accidental, unwanted local failure of the stone during chiselling.

Machine cutting of stone began after the First World War with a pantograph machine and using a particular font designed by McDonald Gill (brother of Erik Gill, whose 1988 book on typography we have referenced). This was followed by computer-controlled 3D, V-cut lettering machines using tungsten carbide cutters. Sand-blasting machines can also be used in which the stone is cut through a stencil. Water-cutting machines also exist and have the advantage of being cooler during the cutting operation. These methods are cheaper

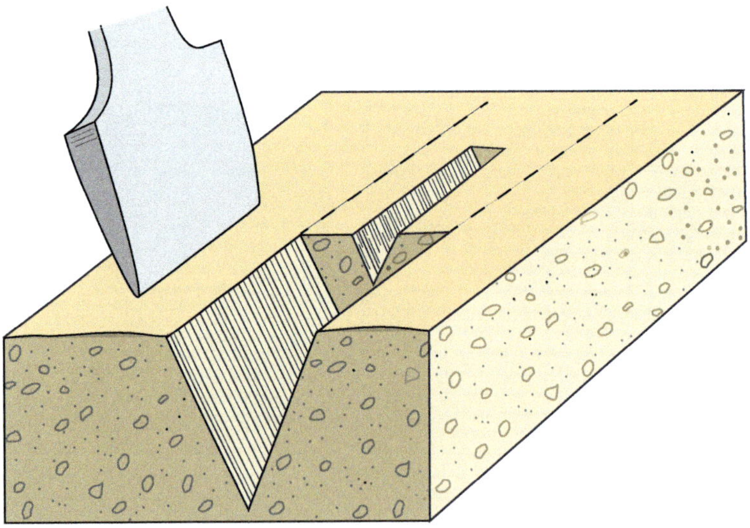

Figure 5.104 Stone cutting so that breakage is towards an existing 'free face', i.e., the earlier cut smaller groove.

but lack the finesse of computer-controlled tungsten carbide cutters, which in turn lack the subtlety and cumulative experience of the skilled and practised stonemason's hand.

An example of the fine detail that can be achieved in granite using machine cutting is illustrated in Figure 5.105. This carved design in red granite is based on the *Ginkgo biloba* tree leaf (with its two lobes, i.e., bi-lobed), the Ginko tree being the official tree of the city of Kumamoto in Japan.

Finally, we note how stone cutting can provide spectacular examples of art in stone walls, the example in Figure 5.106 being in Liverpool Cathedral.

Figure 5.105 Machine carved design in granite, Kumomoto, Japan.

Figure 5.106 The Sheppard Memorial sculpture on an inside wall of Liverpool Cathedral.

NOTES

1 Because some of the lighthouses were built by different generational combinations of the family members, the exact number of lighthouses listed for the family members is subject to discussion.
2 Fractographic, pertaining to fractography: the study of the geometry of fracture surfaces.
3 It is encouraging to know that, during a siege, castle operations were not necessarily conducted by the lord of the castle. For example, at the beginning of the 13th century, it was Gerard de Camville's wife, Nichola, who repelled three attacks on Lincoln Castle during a 25-year period.

Chapter 6

The architecture of stone buildings

Architecture: the art or science of building.

6.1 INTRODUCTION

In his book *An Outline of European Architecture,* Pevsner (1943) explains that the term architecture applies only to buildings designed with a view to aesthetic appeal. So, a bicycle shed, being built solely for practical purposes with no intention of being aesthetically attractive, does not form part of architecture—unlike Lincoln Cathedral. We might also ask about architectural styles and why they developed in the way that they did, especially when stone was used as the main structural building material, from the Egyptians right through to relatively recent times. In a book titled . . . *isms: Understanding Architecture*, Melvin (2005) provides definitions and examples of 55 architectural styles from Pre-Classicism (e.g., the Egyptian pyramids) right through to 'Metarationalism' (i.e., buildings which define a field of spatial experiences overturning conventional structural logic). However, we will confine our discussion to the more mechanical aspects of stone buildings, their relation to architecture and the traditional architectural styles.

The most fundamental evolutionary mechanical division of styles is

1 using stone lintels across two uprights (trilithons) as in Stonehenge,
2 use of the semi-circular arch,
3 use of the pointed arch, and
4 use of modern techniques,

and we will concentrate here on the later styles as related to the early religious stone buildings in Britain: Saxon (i.e., before the 1066 AD Norman Conquest) through Norman, Early English, Decorated and Perpendicular. However, we note the importance of the preceding Egyptian, Greek and Roman styles because features of these styles have permeated right through to the present day, particularly the styles of pillars, which have already been described in Section 5.2.1.

Following the descriptions of these periods, we will note that from a mechanical viewpoint the overall development of stone building architecture was inevitable. We then pass quickly through the Tudor, Elizabethan, Jacobean and Inigo Jones periods and jump to a note on modern architecture and explain the term 'post-modern'. In the history of stone building design, the two architects Marcus Vitruvius Pollio and Sir Christopher Wren are truly exceptional and so we have devoted a section to each. Finally, at the end of this chapter, we list books with glossaries of architectural terms for those wishing to take the subject further.

6.2 AN HISTORICAL NOTE ON THE EARLY USE OF BUILDING STONE IN ENGLAND

Evidence for the earliest use of building stone in England and the associated quarries is hard to come by, but here we provide two examples: one from Roman and one from Saxon times.

Verulamium was a Roman town located at the side of Watling Street, part of which runs from London to Wroxeter in the Welsh borderlands, and which now defines the present route of the A5 trunk road. The town was situated near to the site of St. Albans Abbey, which is 25 miles north of London in Hertfordshire, England. The main development of the town occurred in the second quarter of the second century AD under Hadrian and Pius, and the site was excavated in the 1930s. In an archaeological record of the excavations, the tandem authors Wheeler and Wheeler (1936) note that:

> Freestone[1] is lacking within a distance of 40–50 miles of Verulamium and is proportionately rare there. The columns of the 'triangular' temple were, however, cased in oolite from Ketton in Rutland, whilst the architrave of the south-west gateway was likewise of oolite from Oxfordshire. A few fragments of Purbeck marble were found in and near the 'triangular' temple. The only non-British building material identified at Verulamium is a foreign white marble (probably Carrara) of which fluted and moulded casings were made for some building.

It is quite astonishing to consider the task of transporting heavy marble items from Italy to England in Roman times, much of the journey being over land.

In a paper on The Saxon Building Stone Industry, Jope (1964) notes that in Saxon times, i.e., in the years leading up to 1066 AD, English churches were virtually the only buildings constructed of stone. He reports the results of a survey of Saxon stonework at almost 500 locations in the south of England, see Figure 6.1.

Jope notes that the map, modified here as Figure 6.1,

> emphasizes the importance of the Jurassic oolitic freestones of fine quality, carried 50 miles west into Devon and into the Welsh borderland, 70 miles east into Hampshire and to the lower Thames valley, over East Anglia . . . and along the eastern coasts. All this, however, largely repeats a pattern developed in Roman Britain. The survey shows also the widespread Saxon use of Quarr stone [a freshwater limestone] from the Isle of Wight . . . and its coastal transport eastwards towards the south coast . . ., and perhaps the beginning of the medieval maritime trading in stone from the Bristol Channel area, as well as up the Severn.

Jope also remarks on the fact that "the normal Saxon word for building stone was *stān*[2]" and that "Little evidence about stone quarrying can be derived from Saxon documents."

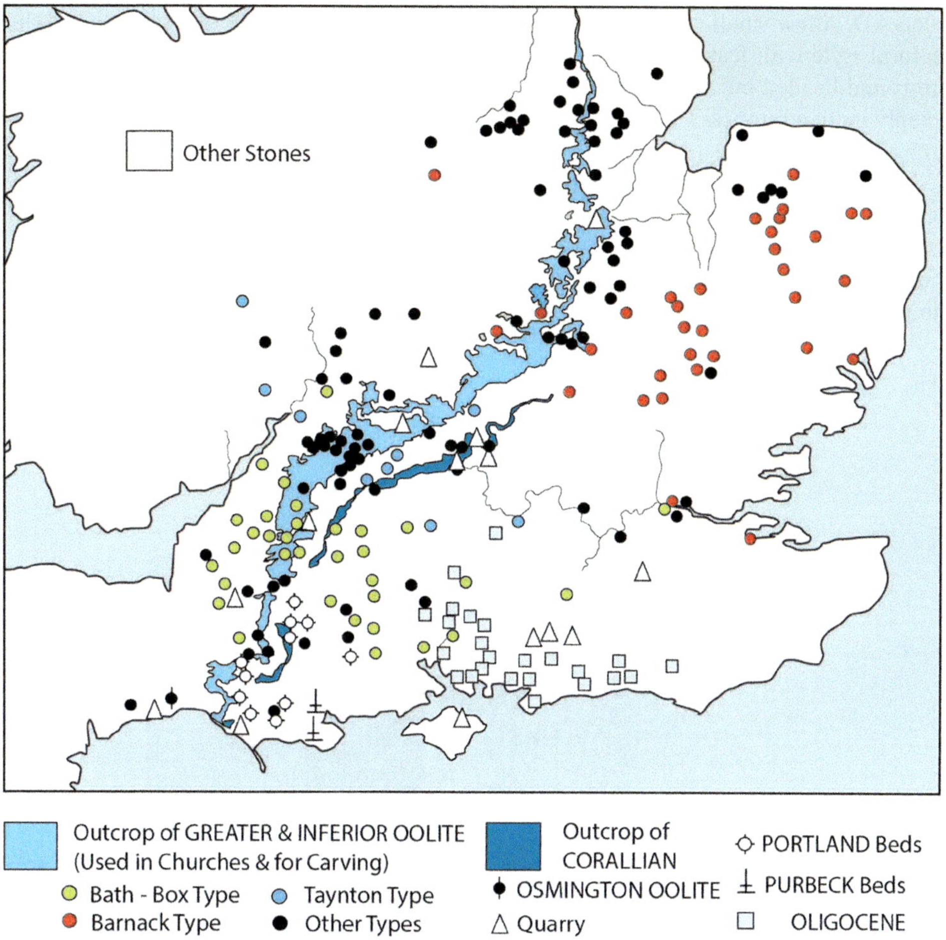

Figure 6.1 The sources of building stone in Southern Saxon England. (Modified from Jope, E. M. (1964). The Saxon building-stone industry in Southern and Midland England. *Medieval Archaeology*, 8, 91–118)

6.3 THE SAXON ARCHITECTURAL STYLE

The Romans occupied Britain from 43–410 AD, leaving many stone structures built in their characteristic style with semi-circular arches as described Section 5.3. But after the Romans' departure, it was to be two centuries of turmoil before the invading Saxons (a group of Germanic tribes originating from near the North Sea coast of present day Germany) converted to Christianity and built stone churches. Over the following centuries, religious buildings were constructed, starting with the Benedictines (656 AD) and followed by the Cistercians, Dominicans, Franciscans and Carmelites. Considering the relatively small population at that time, there was significant religious building activity because of the Christian fear that the world was going to end on the Day of Judgement in 1000 AD.

This Saxon period covers the 600-year period before the Norman Conquest in 1066 AD. A few small churches remain, e.g., Figure 6.2, which are of a primitive architectural style with features such as small windows set high in the walls, some of which are round-headed but some are simply made in triangular form through the use of two steeply inclined stones. The Earls Barton church tower is an exemplary example of Saxon architecture.

The characteristics of Saxon architecture are thick walls of rubble or ragstone, the corners being roughly dressed ashlar (which is masonry made of large square-cut stones used as a facing on walls of brick or stone rubble) set alternately vertically and horizontally, known as long and short work, small windows with round or triangular heads and the tower strengthened and decorated by wall arcading, see Figure 6.2.

Figure 6.2 Saxon church at Earl's Barton, England. (Image from *Cassell's Illustrated History of England*, 1865)

6.4 THE NORMAN ARCHITECTURAL STYLE

The Norman style, based largely on the semi-circular arch, begins the series of evolutionary architectural developments that occurred in Britain from the 11th century to the present. These developments were to a large extent manifested by the many castles and churches built from Norman times onwards, but initially they were not indigenous developments; rather they were the result of the influx of Norman influences following one of the most significant dates in British history, 1066 AD, when, following the Battle of Hastings, William the Conqueror overcame the Saxons and became King of England. The subsequent architectural influence was just part of the Norman customs that then pervaded England and which included language, dress and manners.

Castles and churches in the French style were built to reward the warriors and priests who had helped William to overcome England, see the map in Section 5.4, Figure 5.72, indicating the locations of the English castles. The architectural emphasis was on the semicircular arch and thick, strong walls. In addition to the castles, the churches were potentially subject to attack so the windows tended to be small. One of the exceptions to the overall severity of the style was the emphasis on decoration around the doorways, especially the chevron type patterns. Examples of the way in which the doorways could be enriched through decoration are shown in Figure 6.3(a) and (b).

The Norman architectural style is characterised by thick walls (built with rubble and faced with ashlar—rectangular blocks of masonry), large diameter pillars often with incised geometrical patterns, semi-circular arches, small windows, ornamented doorways and interlaced arcading. In Section 7.3 describing the exemplary Norman/Romanesque style of Durham Cathedral with its majestic incised pillars, we discuss not only the overall geographical defensive location of the cathedral but also its style and the severe weathering of the sandstone used for its construction.

(a) (b)

Figure 6.3 (a) The 'dogtooth' style of Norman arch decoration. (b) The west doorway of St. Mary's church Chepstow, Monmouthshire, Wales, UK.

6.5 THE EARLY ENGLISH ARCHITECTURAL STYLE

Developing from the Norman style, the Early English architectural style represents the beginning of the Gothic period which was a golden age in British architecture and used during the 13th century. The most significant change was from the semi-circular arch to the pointed arch, the latter enabling much better roof vaulting (i.e., the enhanced construction of stone roofs), noting that this was explained in Section 5.3.5. However, increased stone meant a greater weight on the walls which needed buttresses to protect them from bowing outwards and collapsing. Such 'flying buttresses' are easily recognisable on the outside of churches, as are the pinnacles often constructed above them to direct the force vector into a more vertical angle, Figures 6.4(a), 5.35 and 5.36. Additionally, church towers were a feature of this style, being both for decorative and defensive purposes.

Inside the churches, the Early English pillars were more complicated than those of the Norman period, often consisting of a large central shaft surrounded by four smaller shafts, although there were many variations on the theme. The foliated and moulded capitals (the tops of the pillars, often compound) evolved into more complex forms. These developments thus made significant changes from the Norman architectural geometry to both the insides and outsides of the Early English buildings.

(a)

(b)

Figure 6.4 (a) The window and buttress styles of Early English architecture. (b) Salisbury Cathedral (1220–1284).

The Early English style is characterised by walls finished with a parapet, most openings having a pointed arch, long narrow windows, buttresses (often of the flying buttress type), more pillar complexes often with detached columns of Purbeck 'marble', towers and spires and high-pitched roofs. The exemplary building in this style is Salisbury Cathedral in England, Figure 6.4(b).

6.6 THE DECORATED ARCHITECTURAL STYLE

The Early English architectural style evolved into the Decorated style at the end of the 13th century and continued until the arrival of the plague, the Black Death, in 1349 AD. Although, as the period's name implies, it was marked by much increased decorative richness, no new cathedral was begun in this period. The effort was concentrated on the existing churches and superimposing exuberant detail onto the internal style. In particular, windows became larger with more curvilinear shaped bars replacing the earlier linear tracery, noting that this has led to the Decorated style being sub-divided into the Geometrical and Curvilinear periods. The capitals of stone pillars were often carved with representations of natural foliage, see Figure 6.5.

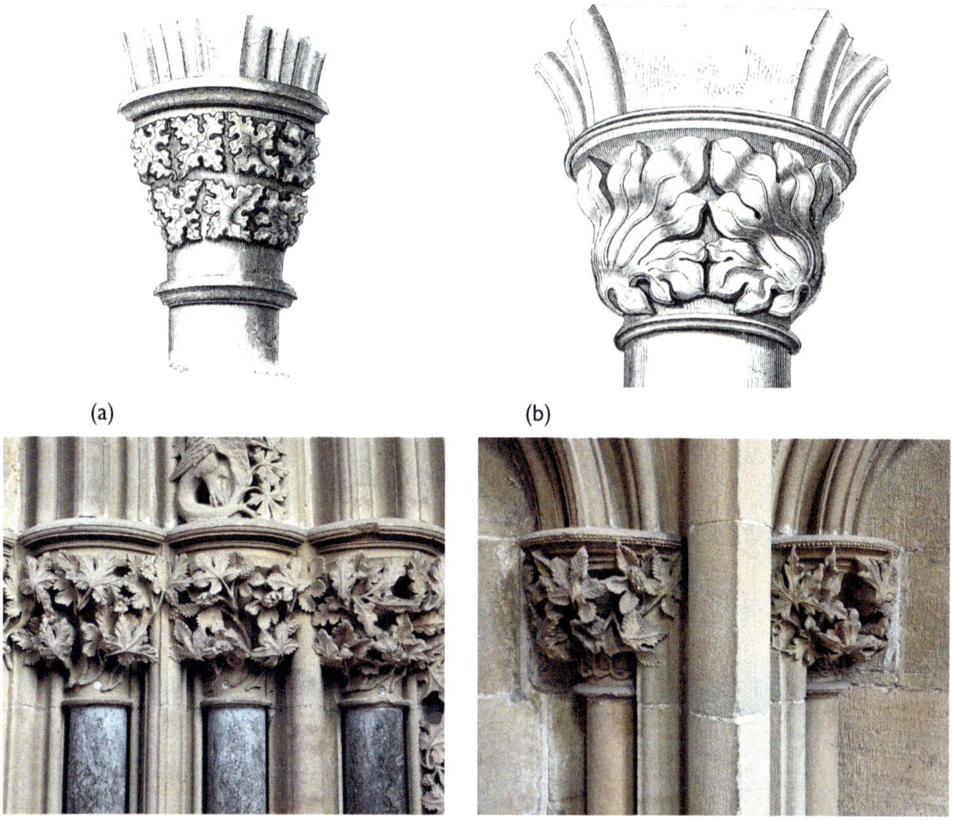

(a) (b) (c) (d)

Figure 6.5 (a) and (b) Pillar capitals in the Decorated style of architecture. (c) and (d) Carved limestone foliage in the Southwell Cathedral Chapter House.

In short, the Decorated period is characterised by richly sculptured doorways, wider windows with elaborate tracery, buttresses with greater projection, enhanced roof vaulting construction and more stone ornamentation. The exquisiteness of the carved stone foliage created during this period is epitomised by the examples shown in the lower part of Figure 6.5 from the Chapter House in Southwell Cathedral near Nottingham in England.

The Decorated style period was an unfortunate time because of the long wars with France and the Black Death, which arrived in 1349 AD. The resultant severe reduction in the population meant that church services and upkeep were significantly reduced. In our companion book on structural geology and rock engineering (Cosgrove and Hudson, 2016), we highlight the case of Norwich Cathedral in England, where construction work on the cloisters was held up by the Black Death for 40 years—during which period the architectural style had changed from the Decorated style to the Perpendicular style, the latter being described in the next section.

6.7 THE PERPENDICULAR ARCHITECTURAL STYLE

From Gloucester church it flew afar The style called Perpendicular.
Thomas Hardy, the son of a stonemason, from
The Abbey Mason (1840–1928)

The Perpendicular style was the last phase of English Gothic architecture[3] and can be divided into two stages: 1350 AD to 1400 AD, when flowing lines 'stiffen' and vertical lines are

(a) (b)

Figure 6.6 (a) The window style of Perpendicular architecture and (b) the interior of Liverpool Cathedral designed by Giles Gilbert Scott[4] in the Gothic Revival style. (Note that the styles of Battersea Power Station, a well-known landmark in London, and the iconic red telephone kiosks were also conceived by him.)

introduced; and 1400 AD to 1450 AD, which is the purer Perpendicular style and is easily recognisable, especially when a Perpendicular type stone screen has been inserted in front of the internal architecture. The other main changes were the increase in the size of the windows and the introduction of fan vaulting in the roof—which is a purely English invention and which we described in Section 5.3.5 and illustrated in Figure 5.69.

The main characteristics of the Perpendicular style of architecture illustrated in Figure 6.6 are high walls, very large windows with through-going mullions (vertical stone divisions between the separate parts of a window), large buttresses and fan vaulting, with Gloucester Cathedral being a superb example of the latter.

* * * * *

All old churches have been renovated over the years, whether in a minor or major way, so when studying such architecture this should be borne in mind. It is rare to find an old building that does not reflect two or more architectural styles. Note the windows in the church at Stow-in-Lindsey, Lincolnshire, England, see Figure 6.7.

Figure 6.7 Windows in St. Mary Minster Church, Stow-in-Lindsey, Lincolnshire. Right-hand window, Saxon; top window, Norman; left-hand window, Gothic.

6.8 SUBSEQUENT ARCHITECTURAL STYLES FOLLOWING THE PERPENDICULAR STYLE

Although there were clear subsequent architectural styles after the Perpendicular period, they do not have as much significance in the context of building stones and stone building because the primary emphasis was then on using bricks as the main building material (noting that we have discussed brick types, brick making and brick bonding in Section 3.9.3). However, many of the elements of the styles already outlined were used in the later Renaissance periods.

- In the **Tudor** period, 1500–1560, the use of red brick with stone was prevalent.
- The **Elizabethan** period, 1560–1600, was characterised by a combination of the earlier Gothic and imported Italian components.
- The **Jacobean** period, 1600–1620, was similar to the Elizabethan period but there was more use of the pillar types, plus the use of the Roman arch and mullioned windows.
- The **Inigo Jones** period, 1620–1660, was when the English Renaissance formal architecture reached its full expression using the Italian style of design. The Inigo Jones work is important to us in the context of building stones and stone buildings because of his use of Portland stone in London.
- The **Wren** Period, 1660–1720, with its classical design, use of pillars, brick and stone together, saw the development of a distinctive style of domestic architecture.
- The **Early Georgian** Period, 1720–1750, when the designs were more academic, emphasising symmetry and proportion, was followed by the **Late Georgian** Period, 1750–1800, when the designs were often based on Ancient Greek architecture.

* * * * *

Architectural styles from the Georgian period up to the more recent post-modern extravaganzas where 'anything goes' will not be described because they have less relevance to the use of stone as a direct building material, although the use of stone in cladding, i.e., in decorating the exterior and interior of buildings, is illustrated in subsequent sections, especially in Section 7.2 on the Albert Memorial in Kensington Gardens in London and in the concluding comments in Chapter 9.

6.9 THE INEVITABLE DEVELOPMENT OF THE ARCHITECTURAL STYLES FROM THE EARLIEST TIMES TO THE GOTHIC

The overall architectural periods from 4000 BC to the present can be considered as belonging to three main types:

- **Trabeated** (from the Latin for beam) architecture, e.g., the Egyptians and Greeks using pairs of columns supporting beams (i.e., refining the style of the Stonehenge trilithons, illustrated in the Frontispiece) during the period, say, 4000 BC to 1 AD.
- **Arcuated** (arc-shaped) architecture, a key feature being the semi-circular arch of the Romans, say, during the period 1 AD to 1200 AD.
- **Gothic** architecture, key features being the pointed arch and ribbed vault, say, during the period 1200 AD to 1500 AD.

Figure 6.8 The architectural evolution of arches.

These primary stages of architectural development were inevitable and caused mainly because of the laws of geometry and physics, plus the human scale. For example and starting at the beginning, a base plus two inclined sides forms a triangle, which is the simplest geometrical closed form and, when extended to three dimensions, is the basis for a wigwam-style enclosure, but this is not easily constructed in stone, nor can it be easily extended laterally. On the other hand, a base, two uprights and a cross piece forms a rectangle, which is not only easy to construct, but also can be extended laterally to a colonnade and rectangular enclosures (Fig. 6.8). Given that the only other geometrically enclosing shapes are more complicated polygons, the development of the simple trabeated style was inevitable.

The development of the arcuated (curved) period is somewhat counter-intuitive because it is based on a curve, which is difficult to generate using stone. However, the breakthrough came when it was realised that a set of stones could be arranged in a vertical plane in a self-supporting geometry not based on pillars and beams—in particular, in a semi-circle (Fig. 6.8). This was indeed a major breakthrough, not only for domestic Roman architecture but also for triumphal arches, for bridge arches, and, by three-dimensional extension, to domes. The natural development from the ability to build a vertical stone semi-circle was to create geometries with more subtly curved components, especially the pointed arch, which led to the Gothic period.

However, after the Gothic period, the architectural styles were sophisticated mixtures of the previous styles, rather than being major geometrical or mechanical breakthroughs, until more recent times when the modern use of concrete, steel and glass has led to different architectural styles and much more complicated and exuberant shapes, although, of course, still constrained by the laws of geometry and mechanics. Thus, from a purely mechanical viewpoint, one can say that the whole architectural evolution has been a circular journey from the Stonehenge-type rectangular openings through to the Perpendicular style rectangular openings. In Britain, this evolution is clearly evident in the churches which have been modified in the different architectural styles: pages of history which he or she who passes the stones may read.

6.10 A NOTE ON MODERN ARCHITECTURE

Another factor related to modern buildings is that it is now possible to create buildings with unusual shapes that are not constructible when stone is used. A recently built example of this is the Canada Water Library (located in the old Surrey Commercial Docks, south of the Thames in London) which has the remarkable outline shown in Figure 6.9. It has an inverted truncated pyramid form clad in aluminium sheets with a relatively small footprint at ground

Figure 6.9 Canada Water Library in Southwark, London.

level, whilst the expanding shape above contains the main library. The motivation for this design was that the floor area required for the library space was larger than the available footprint for the building on the site.

The Canada Water Library is, however, just one of many modern buildings being built around the world. We are living in an 'anything goes' period, as exemplified by the intentional mismatch of the two adjacent buildings shown in Figure 6.10. Since architectural appreciation is highly subjective, we leave it to the reader to decide whether this Montreal example is a stunningly impressive integration of style and stone, a particularly unfortunate example of clashing adjacency, or somewhere in between!

The current trend for spectacular and individualistic buildings is well illustrated by the new recently constructed major buildings in London, but these are now constructed with steel and glass, although stone is often used as a cladding material to enhance the 'architectural authority' of the building. However, there is much discussion about the attractiveness or otherwise of these new buildings, so much so that a "Carbuncle Cup" is now awarded each year to the ugliest new building. This Cup, awarded via the architecture website Building Design, is named after the following comment in a speech made in 1984 by HRH The Prince

Figure 6.10 Mismatch of architectural styles and building stones between adjacent buildings in Montréal, Canada.

of Wales at the 150th anniversary of the Royal Institute of British Architects (RIBA) Royal Gala Evening at Hampton Court Palace, London:

> Instead of designing an extension to the elegant façade of the National Gallery which complements it and continues the concept of columns and domes, it looks as if we may be presented with a kind of municipal fire station, complete with the sort of tower that contains the siren. I would understand better this type of high-tech approach if you demolished the whole of Trafalgar Square and started again with a single architect responsible for the entire layout, but what is proposed is like *a monstrous carbuncle on the face of a much-loved and elegant friend.*

The Carbuncle Cup for 2017 was awarded to the Nova Victoria building, Figure 6.11, which is adjacent to Victoria Station, London, and consists of two 18-storey office buildings and a residential block. The winner of the award each year is decided by a panel of three people with additional comments also taken into account and highlights the subjective nature of observers' reactions, as in the case of the stone mismatch in Figure 6.10. In both

Figure 6.11 The 2017 winner of the Carbuncle Cup for the ugliest building to have been completed in the UK.

of the cases, the Montréal stone mismatch and the Nova Victoria Carbuncle Cup winning building, doubtless there will be those who love and those who hate each example. In the case of the Nova Victoria building, there are some who may not agree with the Carbuncle Cup panel's view and regard the building as imposing and geometrically brilliant, with the red colour introducing dramatic flair.

What is disturbing in the context of this book is that, whilst some of London's old stone architecture is being retained, e.g., the Wren churches, these are now being surrounded by large new majestic glass-fronted office buildings that obscure the sunlight—leaving the old stone smaller buildings in relative darkness. On the larger scale, the same applies to some long appreciated views across London (particularly of St. Paul's Cathedral, which now has 13 protected vistas) that are now being blocked by the new, high office buildings. Examples of the modern buildings are the Shard, 306 m high, and, using their colloquial names, the

Walkie-Talkie building (160 m), the Gherkin (180 m) and the Cheesegrater (224 m), plus many others. There does not seem to be any solution, nor indeed any fully acceptable compromise, to these difficulties as London continues to expand upwards.

6.11 POST-MODERN ARCHITECTURE EXPLAINED: THE CASE OF THE BUNDLED PILASTERS

We conclude this chapter with an explanation of post-modern architecture, because it is a term which can be confusing. Earlier in the book, we included photographs of two post-modern buildings: the 1 Poultry building in London (see Fig. 3.47 in Section 3.5.2) and the National Scottish Museum in Edinburgh (Fig. 3.51, also in Section 3.5.2), which are both clad in sandstone. But what exactly does 'post-modern' mean, and what is the post-modern architectural style? In a book titled *Postmodernism* by Butler (2002), the author chooses the Sainsbury Wing of the National Gallery in Trafalgar Square in London as an illustrative example; the exterior frontage of the Wing being made of the traditional Portland stone, but the architectural style is non-traditional (Fig. 6.12), this being the 1991 revision of the originally proposed design following Prince Charles' adverse comments.

The pilasters referred to in the captions to Figures 6.12 and 6.13 are the vertical pillar-like features which, unlike pillars, are not isolated from the wall and only serve a decorative (rather than structural) purpose. Note that those on the exterior of the Sainsbury

Figure 6.12 The entrance to the Sainsbury Wing extension of the National Gallery in Trafalgar Square, London. Note the asymmetrical 'bundled' pilasters on the right. (A pilaster is a non-supporting decorative pillar connected to the wall of a building.)

Figure 6.13 Symmetrically spaced pilasters on the main building of the National Gallery. Compare the geometrical arrangements of the pilasters in Figures 6.12 and 6.13.

Wing (Fig. 6.12) are 'bundled', rather than being symmetrically positioned, as they are on the adjacent west end of the main National Gallery building (Fig. 6.13). This geometrical asymmetry is typical of post-modern architectural features. Butler (2002) states that "the form-following-function language of modernist architecture was far too puritanical and should allow for the vitality, and no doubt the provocation, to be gained from disunity and contradiction. Work like this happily deconstructs itself." Needless to say, whilst some observers consider post-modern architecture to be stimulating and ironic, others consider it to be at best unnecessary and at worst horrific. As we noted in the previous section, in 1984, Prince Charles compared the originally proposed Sainsbury Wing to be "like a monstrous carbuncle on the face of a much loved and elegant friend." Since then, however, post-modern architecture has been widely accepted.

6.12 TWO EXCEPTIONAL ARCHITECTS: MARCUS VITRUVIUS POLLIO AND SIR CHRISTOPHER WREN

It is outside the scope of this book to describe the lives and works of the many outstanding architects who have designed stone buildings, but we do wish to highlight briefly the major contributions made by Marcus Vitruvius Pollio and Sir Christopher Wren. Vitruvius was a Roman military architect and engineer who lived in the 1st century BC, and Christopher Wren was an English polymath and architect who lived from 1632 to 1723. Both these men made extraordinary contributions to architecture and had specific comments to make on different aspects of building stones and stone buildings—many of which are still relevant today.

6.12.1 Marcus Vitruvius Pollio

Vitruvius wrote the ten book treatise *De Architectura* (On Architecture) in Latin. The main subjects of these 10 books, which are dedicated to his patron, the Emperor Caesar Augustus, are:

- Town planning, architecture and engineering
- Building materials

- Temples and the orders of architecture
- Continuation of temples and the orders of architecture
- Civic buildings
- Domestic buildings
- Pavements and decorative plasterwork
- Water supplies and aqueducts
- Sciences influencing architecture
- Use and construction of machines

In the second book on building materials, Vitruvius covers the origin of buildings, the elements, bricks, sand, lime, pozzolana, stone, different kinds of walls, timber, and highland and lowland fir. It is in this second book that his descriptions and comments on stone and walls are so relevant to our current book and to present day practice. After discussing the bricks, sand, lime and pozzolana (a naturally occurring material, mainly silica, that reacts chemically with slaked lime and moisture, forming a strong cement), Vitruvius notes that stones are all subject to deterioration because of frost and ice and, when near the sea due to salt and "the surges of the sea." He says, with reference to the quarries nearest to Rome that, "Two years before the commencement of the building, the stones should be extracted from the quarries in the summer season: by no means in the winter; and they should then be exposed to the vicissitudes and action of the weather. Those which, after two years' exposure, are injured by the weather, may be used in the foundations; but those which continue sound after this ordeal, will endure in the parts above ground. These rules apply equally to squared as to rubble or unsquared stone work." (translation by Gwilt, 1874). In the context of walls, Vitruvius notes that there are different types of masonry and he discusses their stability and longevity in terms of the stone jointing patterns.

Another interesting comment in Book 2 is as follows:

> The public laws forbid a greater thickness than one foot and a half to be given to walls that abut on a public way, and the other walls, to prevent loss of room, are not built thicker. Now brick walls, unless of the thickness of two or three bricks, at all events of at least one foot and a half, are not fit to carry more than one floor, so that from the great population of the city innumerable houses would be required. Since, therefore, the area it occupies would not in such case contain the number to be accommodated, it is absolutely necessary to gain in height that which could not be obtained on the plan. Thus by means of stones piers or walls of burnt bricks or unsquared stones, which were tied together by the timbers of the several floors, they obtained in the upper storey excellent dining rooms. The Roman people by thus multiplying the number of storeys in their houses are commodiously lodged. Having explained why, on account of the narrowness of the streets in Rome, walls of brick are not allowed in the city. . . .
>
> (translation by Gwilt, 1874)

This discussion by Vitruvius resonates with current discussions on the advantages and disadvantages of the plethora of tall buildings currently being built in London and elsewhere. A more recent translation of Vitruvius' book *On Architecture*, prepared by R. Schofield (Penguin Classics, 2009), is a *tour de force* in exceptionally readable English and is highly recommended for further reading.

6.12.2 Sir Christopher Wren

Christopher Wren lived during an exceptionally disruptive time in the history of England, i.e., during the Civil War (1642–51), the plague (1665–66) and the Great Fire of London (1666). Born in 1632, Wren's early education was from a domestic tutor before his attendance at Westminster School in London starting in 1642, civil war breaking out the same year with its associated major disruptive effects, e.g., when Commissioners had been sent to every county with express orders "for defacing, demolishing, and quite taking away of all images, altars, crucifixes out of all churches and chapels." The house of his father, Dean Wren, at Windsor was sacked in 1645. By 1647, Christopher Wren had invented a weather clock and had written a treatise on spherical trigonometry. In 1649, he entered Wadham College at Oxford University and continued with his inventions. In 1657, when he was 24, he became the Gresham Professor of Astronomy at Oxford, but the civil war continued to disrupt university life, although in 1660 he was able to continue his lectures. Also, in 1660, the Royal Society had its beginnings in Wren's rooms at Gresham College. Then, before he was 30, he was made Doctor of Laws by both Oxford and Cambridge universities.

In the year 1662 and via Charles II, Wren's interests turned from medicine, mathematics and astronomy to architecture when he became Assistant to the Surveyor-General and was required to repair the Palace of Windsor. Also in 1662, he was asked to survey the old St. Paul's Cathedral, which had been severely damaged during the Civil War and, later on, he became heavily involved with the Royal Society. During the plague outbreak in 1665–6, Wren visited Paris, returning in February 1666, and by May had completed a report on the condition of old St. Paul's—which had been used as a cavalry barrack during the Civil War. The four-centuries-old roof had fallen in. Note the following comment that he made in his report:

> First, it is evident by the Ruin of the Roof, that the Work was both ill-design'd and ill-built from the Beginning, ill-design'd because the Architect gave not Butment enough to counterpoise, and resist the Weight of the Roof from spreading the Walls; for, the Eye alone will discover to any man, that those Pillars as vast as they are, even eleven Foot diameter, are bent outwards at least six Inches from their first position; which being done on both sides, it necessarily follows that the whole Roof must first open in large and wide Cracks along by the Walls and Windows, and lastly drop down between the yielding Pillars. [This old St. Paul's Church had been built in 1087–1314.] I cannot propose a better remedy than by cutting off the inner corners of the Cross, to reduce this middle part into a spacious Dome or Rotundo, with a cupola, or hemispherical roof, and upon the Cupola (for the outward ornament) a lantern with a spiring top, to rise proportionately, though not to that unnecessary height of the former Spire of Timber and lead burnt by Lightening.

And so evolved the design of St. Paul's Cathedral as we know it today, Figure 6.14.

Interestingly, Sir Christopher noted that higher buildings were becoming more and more expensive because of the time and labour spent in raising the materials but that this lifting procedure could be improved following techniques that he had observed in France and Italy. He also recommended that the foundation needed replacement. But, on 2 September 1666, five days after the King's Commissioners' visit to St. Paul's with Sir Christopher, the Great Fire broke out, lasted for four days, and destroyed most of London over an area two miles by one mile. Contemporary accounts provide a frightening picture: "The stones of St. Paul's

Figure 6.14 St. Paul's Cathedral, London, the view framed by modern buildings.

flew like Grenades, the melting Lead running down the streets in a Stream and the very Pavements glowing with a fiery Redness . . . the metal belonging to the bells melting." The ruins smouldered for four months, during which time Sir Christopher devised a scheme for rebuilding London, especially St. Paul's and 51 parochial churches.

Pertinent extracts of Sir Christopher's report on the state of St. Paul's after the fire and relating to building stones are as follows:

- It seem'd to have been built out of the Stone of some other ancient Ruines, the Walls being of two sorts of freestone, and those small; and the Coar within was Raggerstone, cast in rough with morter and Putty, which is not a durable way of building.

- A second reason of the Decaies which appeared before the last fire was in probabilitie the former fire, which consumed the whole Roof in the Reign of Q. Elizabeth. The fall of Timber then upon the vault, was certainly one maine cause of the Cracks which appear'd in the Vault and of the spreading out to the Walls above 10 inches in some places from their true Perpendicular as it now appears most manifestly.
- The Portick [the roofed structure leading to the entrance of the Cathedral] is totally depriv'd of that excellent beauty and strength which time alone and weather could have no more overthrown than the naturall Rocks, so great and good were the materials and so skilfully were they lay'd after a true Roman manner. But so impatient is the Portland Stone of fire, that many Tonns of Stone are scaled off and the Columns flaw'd quite through.

Although Sir Christopher recommended a total rebuilding of the Cathedral, on 15 January 1667, King Charles II issued an order requiring restoration of the existing church. Meanwhile, Wren was engaged in a variety of other activities which were discussed by the Royal Society, including experiments for raising weights using gunpowder! However, major problems were encountered by those attempting to restore the stonework of St. Paul's leading, eventually to letters patent under the Great Seal being issued on 12 November 1673 authorising complete rebuilding of the Cathedral—and the final result is clearly visible to us today. To assist in the building, and in case something adverse happened to Sir Christopher in the meantime, a 1:25 scale model was made in 1673–4 (by William Cleere) of the future Cathedral using oak, plaster and lime wood, although some changes to this original design were made later. Gunpowder was used to remove the old building and 47,000 loads of rubbish were removed. Note that, although the present St. Paul's Cathedral is large at 515 ft in length, the old St. Paul's was even larger at 586 ft in length. Problems encountered in the rebuilding process were related to the quality of the ground below the cathedral and 'the ungodly custom of swearing among the labourers'.

Sir Christopher Wren's name is primarily associated with St. Paul's Cathedral (and readers are encouraged to visit it); however, he was associated with many other projects across England, not least of which was the Monument in London and the rebuilding of 51 of the 88 parish churches damaged by the Great Fire. His work as Surveyor-General to the King from 1671–7 on the design and building of the Monument recording the Great Fire was in association with Dr Robert Hooke, the City Surveyor, whom we have earlier highlighted in connection with the design of catenary-based arches. The Monument is described in Section 5.2.1 on pillars and shown in Figure 5.5.

If ever there was a good example of the Biblical saying "Cometh the hour, cometh the man", it is Sir Christopher's work following the Great Fire of London. Between 1663 and 1706, he was involved in 107 architectural projects, mainly in London (which included the upgrading of 51 of the 88 parish churches that had been damaged by the Great Fire) and Cambridge, Oxford, Winchester, Chichester and Rochester. He designed and oversaw the building (1672–87) of the St. Stephen Walbrook Church in the city of London. Figures 6.15, 6.16 and 6.17, respectively, show the interior of the church, its dome, and a close up of the travertine stone used for the central altar. The Figure 6.15 view of the interior architecture with its Corinthian columns and the dome in Figure 6.16 is reminiscent of St. Paul's Cathedral architecture, albeit on a much simplified and reduced scale; indeed, it is thought to have been a 1672 'trial' dome for St. Paul's Cathedral.

Figure 6.15 The interior of St. Stephen Walbrook Church, London. Note the characteristic right-angled pillar support geometry and the relatively recent circular travertine altar designed by the sculptor Henry Moore.

Figure 6.16 The dome of St. Stephen Walbrook Church, London.

Figure 6.17 Close up of the modern travertine altar stone in St. Stephen Walbrook Church, London.

The church was damaged during the Second World War and, as part of the upgrade, Henry Moore, the sculptor, designed the present central altar, Figures 6.15 and 6.17, made from Italian travertine—which is a concretionary limestone formed by chemical precipitation of calcium carbonate (calcite) from hot hydrothermal spring waters and is often used as a 'prestige stone', e.g., also as the altar in St. Martin-in-the-Fields Church, Figure 6.18, and for stone benches in the British Library, both in London. Needless to say, after the restoration of St. Stephen Walbrook Church, the central circular travertine altar caused considerable controversy, which was only resolved when the Moore altar was ruled acceptable for the Church of England by the highest ecclesiastical court of the land, the Court of Ecclesiastical Cases Reserved.

To illustrate some of Sir Christopher's thinking 'outside the box', we quote the following extract passages from his letter to a friend in 1708 concerning the Act of Parliament passed to erect 50 new additional parish churches in the City of London and Westminster.

1 First, I conceive the Churches should be built, not where vacant ground may be cheapest purchased in the Extremities of the Suburbs, but among the thicker Inhabitants, for Convenience of the better sort, although the Site of them should cost more: the better Inhabitants contributing most to the future repairs, and the Ministers and Officers of the Church, and Charges of the Parish.
2 I could wish that all burials in Churches might be disallowed, which is not only unwholesome, but the Pavements can never be kept even, nor Pews upright: and if the Churchyard be close about the Church, this is also inconvenient, because the ground being continually raised by the Graves, occasions, in Time, a Descent into the Church, which renders it damp, and the Walls green, as appears evidently in all old Churches.

Figure 6.18 Travertine altar in St. Martin-in-the-Fields Church, Trafalgar Square, London.

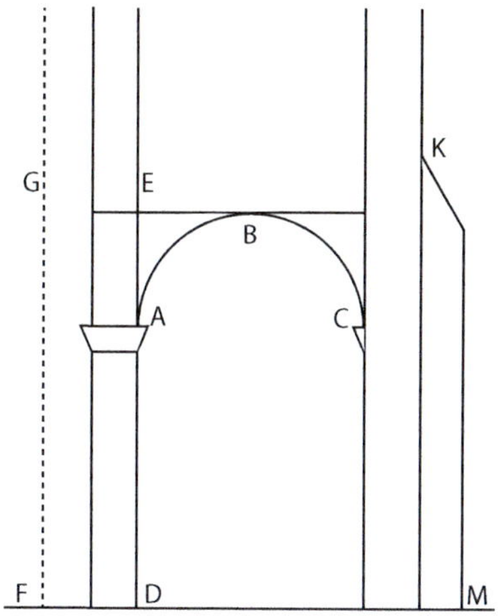

Figure 6.19 Sketch by Sir Christopher Wren in 1713 (redrawn) concerning arch stability, see text.

Also, it is evident that he had a clear understanding of arch stability, as demonstrated by his Figure 6.19 diagram and the following text from a memorial to the Bishop of Rochester in 1713 concerning the Abbey Church of St. Peter at Westminster.

> Let A B C be an Arch resting at C, against an immoveable Wall K M, but at A upon a pillar A D, so small as to be unable to be a sufficient Butment to the Pressure of the Arch A B: what is to be done? I cannot add F G to it to make a Butment, but I build up E so high, as by Addition of Weight, to establish it so firm, as if I had annexed FG to it to make a Butment.

This explanation by Sir Christopher anticipates exactly the modern vector analysis understanding that we outlined in Section 5.3 and Figures 5.35 and 5.36.

6.13 BOOKS CONTAINING EXPLANATIONS AND GLOSSARIES OF ARCHITECTURAL TERMS

As the reader will have noted in the previous sections of this chapter, there is a variety of terms used in the architectural context to describe the components of stone buildings; some of these terms are in common use, others may be unclear, and indeed some others may be unfathomable. The task of explaining all these architectural terms is a large one and beyond the scope of this book. However, over the years, whole books have been dedicated to this subject, some of which (from 1850–2014) are highlighted in the following list—presented in alphabetical order of the first author.

- Child, M. (1981) *English Church Architecture: A Visual Guide*. London: B.T. Batsford Ltd., 119p.
- Cole, E. (ed.) (2002) *The Grammar of Architecture*. London: Bulfinch Press, Little, Brown and Co., 352p. Well-illustrated and explained double-page spreads on the components of the architectural periods.
- Cragoe, C.D. (2014) *How to Read Buildings: A Crash Couse in Architecture*. London: Bloomsbury Publishing, 256p. Pocketbook-sized guide to decoding a building's style, history and evolution.
- Fletcher, B. (e.g., 1943) *A History of Architecture on the Comparative Method*. B.T. Batsford Ltd., London, 1033p. Outstanding and extensive compilation of architectural information and supporting diagrams. 11th edition. The first edition was in 1896; many editions have followed.
- Harris, J. and Lever, J. (1966) *Illustrated Glossary of Architecture 850–1830*. London: Faber and Faber, 302p. (of which 224 are full page photographs).
- Hopkins, O. (2012) *Reading Architecture: A Visual Lexicon*. London: Lawrence King Publishing Ltd., 174p. The word 'lexicon' means the vocabulary of a branch of knowledge. The technique used in this book is to illustrate a particular topic and include clarifying text around the detailed sketches and photographs.
- Jenner, M. (1993) *The Architectural Heritage of Britain and Ireland: An Illustrated A–Z of Terms and Styles*. London: Michael Joseph, 320p. A well-illustrated glossary of architectural terms.
- Melvin, J. (2005) *. . . isms: Understanding Architecture*. London: Herbert Press, 160p. Two-page spreads explaining 54 architectural styles.
- Parker, J.H. (1850) *A Glossary of Terms Used In Grecian, Roman, Italian and Gothic Architecture*. 5th edition. Enlarged with 1700 woodcuts. 3 vols., London: David Bogue, 528p.
- Parker, J. H. (1875/1986). *Classic Dictionary of Architecture: A Concise Glossary of Terms Grecian, Roman, Italian and Gothic Architecture*. 4th edition revised. London: Cassell plc., New Orchard Editions (facsimile) 327p. Distributed in the USA by Sterling Publishing. Illustrated re-published Victorian book with excellent explanations.
- Parker, J. H. (1888) *A Concise Glossary of Terms used in Grecian, Roman, Italian and Gothic Architecture*. London: Parker and Co., 335p.

- Parker, J. H. (1898) *ABC of Gothic Architecture*. 10th edition, London: James Parker and Co., 265p.
- *Pevsner's Architectural Glossary* (2016) London: Yale University Press, 144 pages. A most helpful guide, pocketbook size, well-illustrated, colour pictures. *This glossary is recommended as being the most recent and convenient to use.*
- Rice, M. (2009) *Rice's Architectural Primer*. London: Bloomsbury Publishing, 240p. Some text but mostly delightful coloured sketches of architectural terms and buildings.
- Rice, M. (2013) *Rice's Church Primer*. London: Bloomsbury Publishing, 224p. Attractive hand-drawn sketches of buildings and their components.
- Rutter, F. (1923) *The Poetry of Architecture*. London: Hodder and Stoughton Ltd., 189p.
- Stratton, A. (1937) *The Styles of English Architecture. Part I: The Middle Ages*. London: B.T. Batsford Ltd., 35p.
- Wynne, G. (1930) *Architecture*. London: Thomas Nelson and Sons, 131p.

NOTES

1 Freestone: a fine-grained sandstone or limestone that breaks freely and can be cut and dressed easily in any direction.

2 The Old English word *stān*, meaning building stone, has been the basis for many English place names, e.g., Staines and Stanwell in Middlesex. There are at least 25 Stantons.

3 The English Gothic style of architecture began in the late 12th century and finished in the mid-15th century.

4 Sir Giles Gilbert Scott was the grandson of Sir George Gilbert Scott, who was champion of the Gothic revival and designed St. Pancras Station (Figure 3.103) and the Albert Memorial (Section 7.2).

Chapter 7

Two exemplary stone structures

The Albert Memorial and Durham Cathedral

7.1 INTRODUCTION

Following the structure of this book, i.e., presenting firstly the building stones and then the stone buildings, in this chapter we now highlight two outstanding stone structures: the **Albert Memorial** and **Durham Cathedral**, both in Britain. The first exemplar is a superb *decorative building stone* structure with its spectacular architectural design and abundance of stone types, but we have also included explanations of its structural components in order that the full nature of the Memorial can be understood. The second exemplar is chosen to illustrate a masterly *stone building* structure from the 12th century with its majestic pillars and innovative roof vaulting, but with the difficult problem of sandstone weathering.

7.2 THE ALBERT MEMORIAL

7.2.1 Introduction and background

Prince Albert, the Prince Consort to Queen Victoria, organised with Henry Cole the Great Exhibition held in London in 1851. This was the first international exhibition of manufactured products and was held in the purpose-built Crystal Palace in Hyde Park. The exhibition was a resounding success and it was then that the public, finally and rather begrudgingly, took Albert to their heart. However, 10 years later, Prince Albert died from typhoid at the untimely age of 42. It was then agreed that an appropriate memorial to him should be constructed. Several notable architects were invited to submit plans for the memorial and those by George Gilbert Scott (1811–78) were chosen. He was a prolific English Gothic revival architect, having been inspired by August Pugin (1812–52), whose own work had culminated in the interior design of the Palace of Westminster. Scott was later knighted as a reward for the design and construction of the memorial.

Scott's memorial was designed to combine the skills of the jeweller, enamellist and artist with those of the architect. The style he chose was that of the 13th century, i.e., the high and late medieval period, the age he felt was the finest period of indigenous architecture of the countries of northern Europe. An entry in his note book for 10 March 1864 reported in his *Personal and Professional Recollections* (1879) reads:

> My idea in designing it was, to erect a kind of ciborium [a canopy over an altar in a church, standing on four pillars] to protect a statue of the Prince; and its special characteristic was that the ciborium was designed in some degree on the principles of the ancient shrines. These shrines were models of imaginary buildings, such as had never

> in reality been erected; and my idea was to realise one of these imaginary structures with its precious metals, its inlaying, its enamels etc., etc. This was an idea so new, as to provide much opposition.

Thus, Albert's memorial was to be Scott's own enlarged, modern version of a 13th-century shrine, enriched by emblematic statuary and by all the decorative arts in conjunction: it was to be a summary of 19th century aspirations, Figure 7.1 and 7.2(a).

In an outline description given by Scott relating to his design of the memorial, reported in the *Handbook to the Prince Consort National Memorial* (Scott, 1874), he states:

> The idea which I have worked out may be described as a colossal statue of the Prince, placed beneath a vast and magnificent shrine or tabernacle, and surrounded by works of sculpture illustrating those arts and sciences which he fostered and the great undertakings which he originated. I have in the first place elevated the monument upon a lofty and wide spreading pyramid of steps. From the upper platform rises a podium or continuous pedestal, surrounded by sculptures in *alto-rilievo* [high relief extending out from the background to more than half their depth], representing historical groups or series of the most eminent artists of all ages of the world, the four sides being devoted severally to Painting, Sculpture, Architecture, Poetry and Music. . . . This forms, as it

Figure 7.1 The monument situated on the edge of an incised river terrace of the Thames. The sunlit front of the monument faces south.

> were, the foundation of the monument, and upon it is placed the shrine or tabernacle already mentioned. This is supported at each of its angles by groups of four pillars of polished granite, bearing the four main arches of the shrine. Each side is ornamented by a gable, the tympanum [wall surface over an opening] which contains a large picture in mosaic and inlaid with mosaic-work, enamel and polished gem-like stones; thus carrying out the characteristics of a shrine.

Scott also states that, "There can, indeed, be no doubt that the public expect a monument of great and conspicuous magnificence . . . and . . . the structure is to have a shrine-like appearance, and be enriched to the utmost extent all the arts can go." Twenty years prior to the building of the memorial, the architects Barry and Pugin had completed the Palace of Westminster, i.e., the Houses of Parliament, which is the epitome of the Gothic revival style but, by the time the memorial was constructed, tastes had begun to change. As a result and from the memorial's inception to the present day, it has been subjected to both criticism and abuse. As Robinson (1987) notes in the opening sentence of his paper on the geology of the Memorial, "There can be no half measures to opinions about the Albert Memorial: you either like it or loathe it."

Before his death and in connection with the potential erection of a statue to himself in Hyde Park as part of a memorial to the 1851 Exhibition, Prince Albert had commented that he disliked the idea and he wrote, "I can say, with perfect absence of humbug, that I would much rather not be made the prominent feature of such a monument, as it would both disturb my quiet rides in Rotten Row to see my own face staring at me, and if (as is very likely) it became an artistic monstrosity, like most of our monuments, it would upset my equanimity to be permanently ridiculed and laughed at in effigy." This quote is taken from Sheppard (1975), who adds that, "The story of the memorial that posthumously realised much of the Prince's foreboding is one well documented in some of its aspects, but not perhaps greatly illuminating the central fact of an artistry inadequate to the high and difficult aim of the designer."

Regardless of the controversy, the Albert Memorial represents the nation's greatest monument to the Prince Consort. It is a spectacular masterpiece of Gothic revival architecture, a bejewelled shrine which symbolises Queen Victoria's grief and a nation's gratitude to an extraordinary man. In addition, and in the context of this book, it represents a plethora of structural and decorative stones gathered from all corners of the British Isles. The sketch, provided as Figure 7.2, indicates the various sections of the monument; these are discussed in the text, together with some of the building stones used in its construction.

7.2.2 Foundations and undercroft

Robinson (1987) points out that the monument is situated on the edge of an incised river terrace (Fig. 7.1), an ancient flood plain of the Thames deposited some 100,000–200,000 years ago in the Late Pleistocene period during a warm, inter-glacial phase. Flood waters from a later glacial period cut a channel into the terrace and the drop in topography can be clearly seen in Figures 7.1 and 7.3, which show the monument located on the terrace edge, and the carriageway in front at a lower level occupying an area where part of the terrace was removed by the river. The rise from this excavated region onto the terrace edge (known by geomorphologists as the 'riser') is approximately 3 m, and immediately in front of the monument can be ascended by two flights of six granite steps, separated by a paved platform.

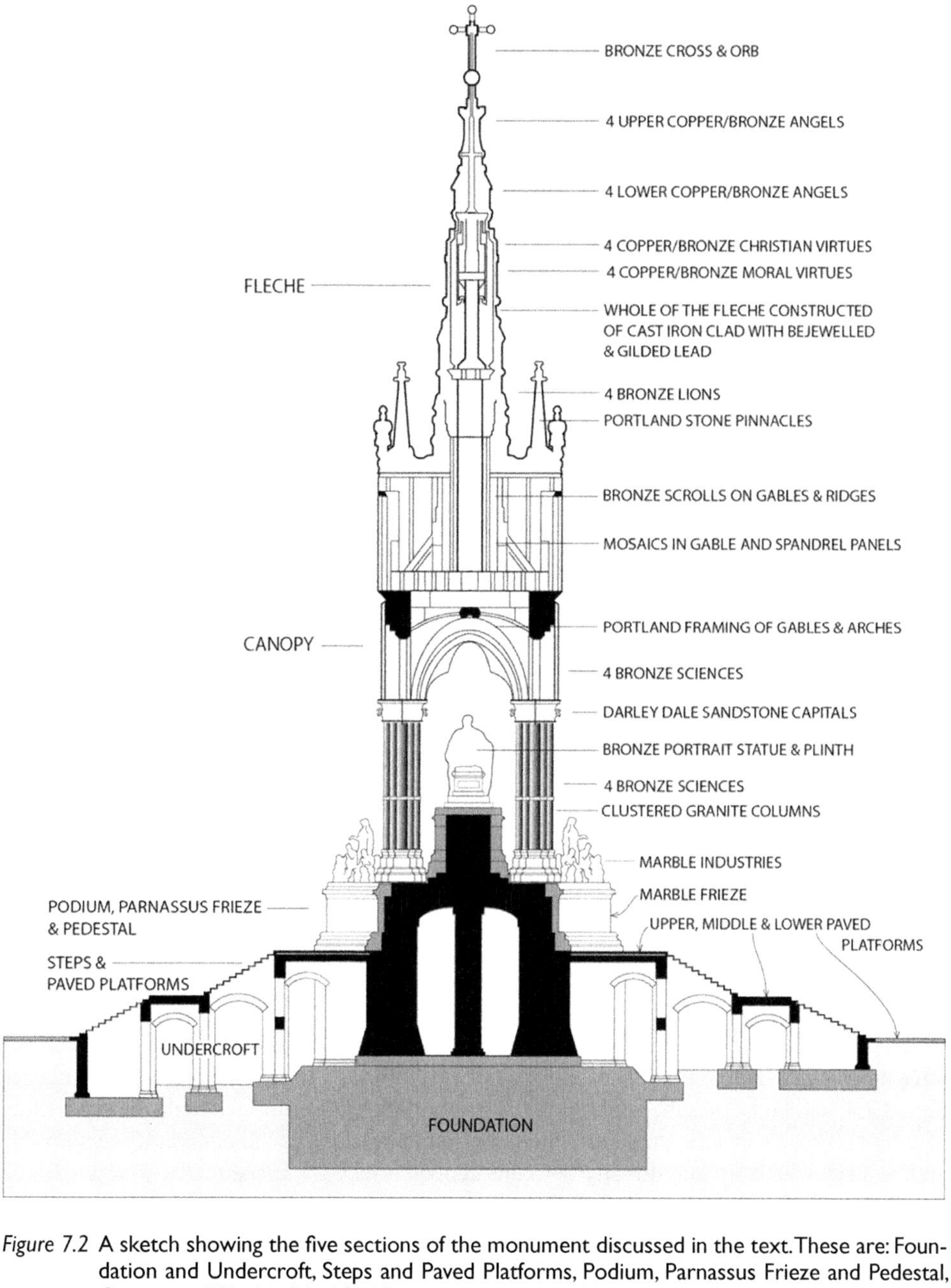

Figure 7.2 A sketch showing the five sections of the monument discussed in the text. These are: Foundation and Undercroft, Steps and Paved Platforms, Podium, Parnassus Frieze and Pedestal, Canopy and Fleche. (Modified from an engraving in *The National Memorial to His Royal Highness the Prince Consort*, 1873, Kensington Central Library)

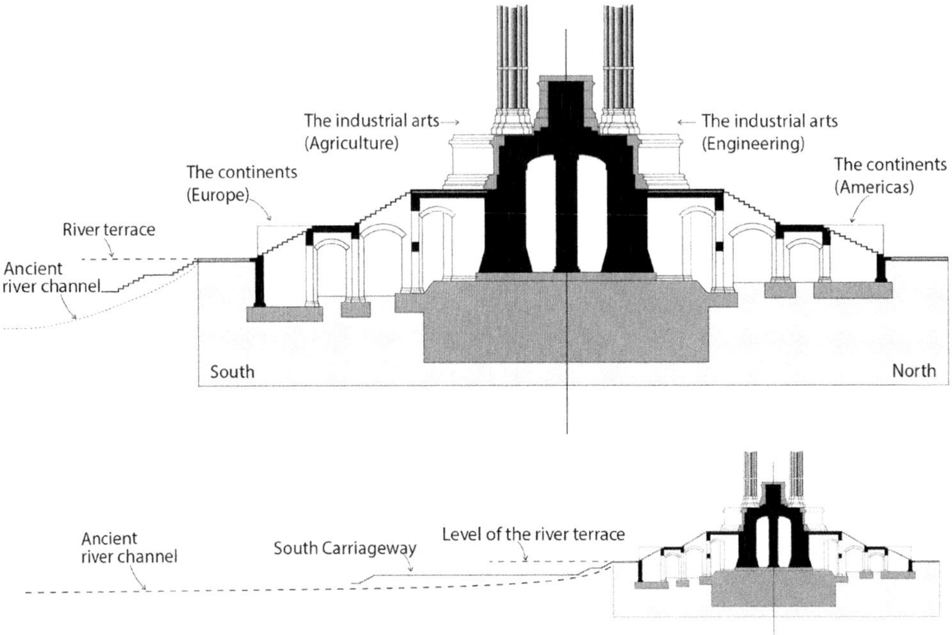

Figure 7.3 Upper diagram: a section, looking west, through the foundation of the monument showing a central concrete slab directly beneath the structure and an extended series of brick arches (the undercroft) above and around it. Lower diagram: the position of the memorial on the edge of a river terrace of the Thames. An ancient river channel running from west to east cuts into the terrace immediately to the south of the memorial. (Modified from an engraving in *The National Memorial to His Royal Highness the Prince Consort*, 1873, Kensington Central Library)

The main foundation of the monument was dug at the edge of the terrace and comprises a large slab of concrete some 17 ft (~5 m) deep seated on river gravels, Figure 7.3. Constructed around and above this foundation is a large complex of brick arches, the 'undercroft', see Figures 7.3–7.5, noting that an undercroft is traditionally a cellar or storage room, often brick-lined and vaulted. The brickwork is comprised of 365 piers and 868 arches, Figure 7.4, and buttresses the central column supporting the memorial and the heavy weight of John H. Foley's bronze statue of Albert. The outer portions of the undercroft support the granite steps which are laid directly onto the stepped brickwork piers, Figures 7.3 and 7.4.

The bricks are London Stock Brick, whose yellow colour and soft appearance come from the yellow local clay, i.e., the London clay of the Lower Eocene epoch, deposited about 56–48 million years ago, from which the bricks were made. They were supplied by Messrs. Richardson, whose works at Acton were within easy distance of the site. The foreman bricklayer, William Jacobs, had an excellent pedigree, having worked on the Houses of Parliament and Nelson's Column. Building of the monument started in May 1864 and, by the end of the year, the foundation and the brick sub-structure had been completed and were greatly admired.

Figure 7.4 A series of brick arch supports (the 'undercroft') built above and around the main foundation. Note the stepped nature of the upper surface of the supports (piers) onto which the granite steps of the basal pyramid would be laid. In the middle part of the figure, the brick core of the podium (see Fig. 7.3) has been clad and the as yet uncarved marble face can be seen. The pedestal is clad in granite and marble. (Contemporary photograph, about 1866, © Kensington Central Library, London)

7.2.3 Steps (structural features)

As noted earlier, in Scott's description of his design, he states that, "I have in the first place elevated the monument upon a lofty and wide spreading pyramid of steps." These steps rise from a lower, paved platform and are broken at the halfway point of their ascent by another paved platform. As can be seen from the upper half of Figure 7.5, the lower flight consists of a square-based pyramid and the upper flight of an octagonal-based pyramid. A variety of granites has been used in the construction of the memorial, the majority of which was used for the steps. The granite used for the three flights of steps leading up from the main road to the terrace on which the monument is built, Figure 7.3(b), was obtained from the Penryn quarry near Falmouth in Cornwall, Britain. The quarry is situated in the Carnmenellis granite, one of a series of granite masses which include the Dartmoor and Bodmin granites

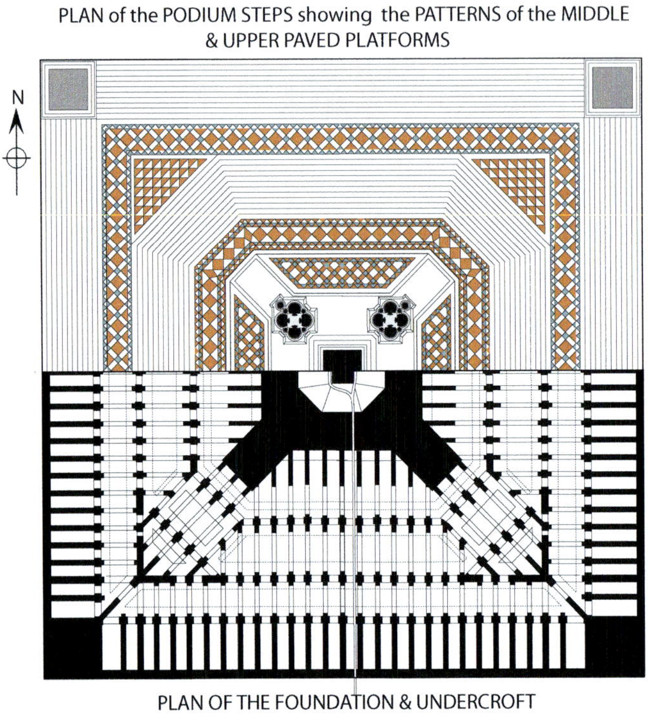

Figure 7.5 Ground plan of the memorial showing, in the upper part, the pyramid of steps on which the structure is built and the designs for the paved areas at mid-height and at the top of the stairs. In the lower section, the detailed layout of the underlying brick undercroft is shown, see Figure 7.4. (Modified from an engraving in *The National Memorial to His Royal Highness the Prince Consort*, 1873, Kensington Central Library)

and which define the spine of the Cornubian peninsula., see Figure 2.14 in Section 2.7 in Chapter 2. They are Late Carboniferous–Early Permian in age (~310–290 million years ago) and intruded into folded Carboniferous and Devonian rocks. The granite is a typical grey Cornish granite containing an interlocking mesh of relatively large, rectangular laths (long and thin crystals) of white, plagioclase feldspar set in a matrix of finer grained grey to glassy quartz and black, biotite mica.

In order to reach the base of the podium from the lower terrace, it is necessary to ascend the two flights of 12 steps shown in the upper part of Figure 7.5, made of a different 'granite', namely 'Castlewellan granite', which has a sombre, grey-blue colour, white feldspar laths (long and thin), less quartz than the Cornish Penryn granite, and containing more dark (ferro-magnesian) minerals; this mineral assemblage indicates that the rock is better described as a granodiorite, see Section 3.2 on granites. It is of Devonian age (400 million years), i.e., older than the Penryn granite and was quarried from the Newry 'granite' of County Down in Northern Ireland. It is described by Kinahan (1888) in his paper on Irish granites as "a peculiar variety of Newry Granite; the black graining of the grey being in oblique lines, giving the stone a unique and chaste aspect."

7.2.4 Paved platforms (decorative features)

The memorial contains three paved platforms separated by steps. There is also a paved area separating two flights of steps that decend from the memorial to the southern carriageway, Figure 7.6. The lowest and largest terrace, Figures 7.7 and 7.9a, surrounds the base of the pyramid of steps on which the memorial stands. Another is situated half way

Figure 7.6 The paved terrace in the flight of steps descending from the lower terrace to the south carriageway. The steps are of Cornish granite from the Penryn quarry in the Carnmenellis granite, see Figure 2.14.

Figure 7.7 Detail of the paving of the lower terrace looking south towards the south carriageway. Note the uneven wearing of the different rock types, a consequence of them being selected primarily for their decorative qualities rather than their strength.

Figure 7.8 The upper flight of stairs, made of Castlewellan 'granite', which separate the middle and upper paved platforms, see Figure 7.5.

up the steps and separates the lower, square-based pyramid from the upper octagonal pyramid, Figures 7.5 and 7.9b. The highest platform is at the top of the steps, Figures 7.5, 7.8 and 7.10.

In addition to the granite that was used to frame the paved areas, four distinct rock types were used for the paving slabs. These are:

- Darley Dale stone
- Hopton Wood stone
- Charnwood Forest slate
- New Red sandstone

All these are fine-grained, originally sedimentary rocks and were selected for their contrasting colours. Figure 7.9 shows a detail of the lower platform which contains examples of the four paving stones listed above. Darley Dale or Stancliffe sandstone, quarried in Darley Dale in Derbyshire, is a yellow–brown stone belonging to the Millstone Grit series of mid-Carboniferous times, about 320 million years ago. It was deposited during the build-up of deltas as rivers entered a sea and has been used in a variety of civic buildings in northern England, including the columns of Leeds Town Hall and the whole of the front of St. George's Hall in Liverpool. Hopton Wood stone forms a particular horizon within the Carboniferous limestone and is quarried west of Middleton-by-Wirksworth, also in Derbyshire. It is white and described as 'a very fine limestone, almost like marble' and as 'England's premier decorative stone'. It is particularly suited for sculpting; Henry Moore used the stone extensively, and it is popular for tombstones, including many thousands for the Commonwealth War Graves Commission. The red paving stones of the pavement are of New Red

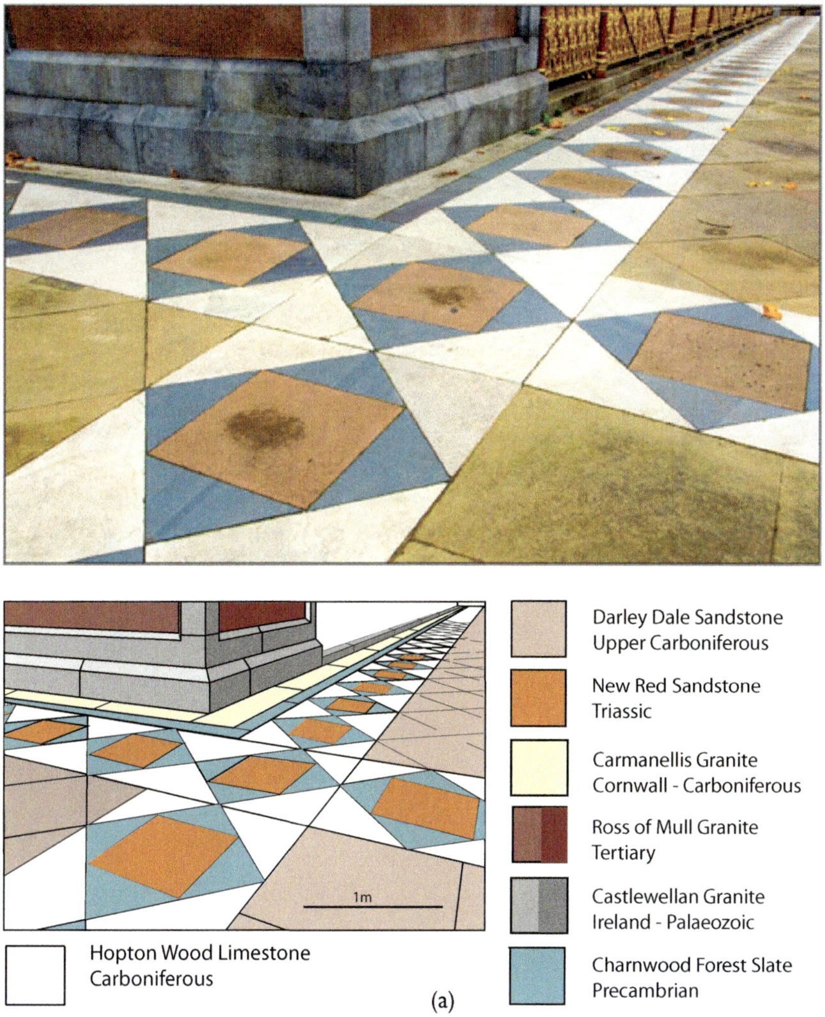

(a)

Figure 7.9(a) Detail of the lower platform and one of the granite pedestals with a key identifying the main rock types.

(b)

Figure 7.9(b) Detail of the middle paved platform, see Figure 7.5, showing the different rock types used in its construction.

Figure 7.10 The upper pavement from which the memorial rises. The paving stones are New Red sandstone (Permian), white Hopton Wood limestone (Carboniferous) and grey-blue Charnwood Forest slate (Pre-Cambrian). The steps are of Castlewellan granite, the base of the podium is of the red Ross of Mull granite, and the carved Parnassus frieze of Campanella marble.

sandstone quarried in Mansfield in Nottinghamshire. This rock is Permo-Triassic in age and red because it was deposited under arid conditions when Britain was situated further south in the desert latitudes some 280–200 million years ago. In such an environment, iron is oxidised to ferric oxide, which has a characteristic red colour. The final member of the paving stones is the grey-blue Charnwood Forest slate. This rock was quarried from the Pre-Cambrian massive in Leicestershire and was deposited ~600 million years ago as a fine-grained volcanic ash. During the tectonic plate motions linked to the closing of the Pre-Cambrian Iapetus Ocean, it was subsequently metamorphosed (heated and deformed) into a slate, making it ideal for paving.

By using paving stones of different shapes (triangles, squares, rectangles and other quadrilaterals) and by exploiting their distinctive and contrasting colours, it was possible to produce elaborate and impressive patterns, e.g., Figure 7.5, middle terrace, and Figure 7.10, that are in keeping with the style of the memorial. The paved areas are framed with granite edgings or steps, Figure 7.8; the crisp outline of these blocks, plus the preservation of the texture tooled onto the dressed surfaces to prevent slipping almost 150 years ago, attest to the resilience of this stone when subjected to weathering and wear. Unfortunately, the same cannot be said of the sedimentary stones making up the paved platforms because they do not have the strength or resistance to weathering of the granites. In addition, the differences in their resistance to weathering and wear has produced irregularities in the platform surface as, for example, where the relatively soft New Red sandstone forms a negative feature where it abuts against the more resistant Charnwood Forest slate.

7.2.5 Podium and pedestal (mainly structural) and frieze (decorative)

The pyramid of steps at the base of the memorial is capped by a podium which supports the canopy and the pedestal on which rests the monumental bronze statue of Prince Albert. As can be seen from Figure 7.3, the central brick columns of the undercroft form the core of the podium and pedestal and this was subsequently clad in a variety of stones.

The upper and lower portions of the podium are clad in Ross of Mull granite, Figures 7.8 and 7.11, obtained from quarries on the westernmost tip of Mull facing the Island of Iona and belonging to the Duke of Argyll. Its pink colour is the result of the feldspars being orthoclase, i.e., the potassium-rich variety as opposed to the white sodium and calcium dominated plagioclase feldspars that characterise the Penryn and Castlewellan granites used for the steps. The first blocks of Ross of Mull granite for the lower part of the podium were emplaced in November 1865. The remaining blocks and the blocks of marble for the magnificent Parnassus frieze which encircles the podium, Figure 7.11, were emplaced over the next six months.

The stones used for the construction of the Albert Memorial are from different parts of the UK, arguably to symbolise the whole kingdom playing a role in the construction of the Monument. The principal exception is the Campanella marble used for the podium frieze, Figure 7.11. For the frieze sculptures, it was originally proposed to use Carrara marble, an Early Jurassic limestone which was deformed and metamorphosed to marble during the Alpine Orogeny. It is excavated from several sites in the Massa, Carrara and Serravezza region of Tuscany (see Section 4.3.3). However, because of the polluted air in Victorian London, the similar but more hard-wearing variety of Carrara marble called Campanella marble was chosen; it was also used for the eight principal sculptural groups (the four continents and the four industrial arts), Figures 7.1, 7.3, 7.8 and 7.14, despite it being more difficult to carve. The name of the Campanella marble originates from its ability to ring when struck with a hammer—a property which visitors to the monument are discouraged from confirming.

The podium is extended at its corners to provide pedestals for the four sculptural groups entitled the 'Industrial Arts', Figures 7.2, 7.8 and 7.14, and this extends the length of the frieze. It has a total length of approximately 64 m (210 ft) and depicts the greatest artists, painters, musicians, writers and architects in Western history. The frieze was intended by the architect Scott

Figure 7.11(a) Part of the poets' and musicians' frieze, the work of Henry Hugh Armstead.

Figure 7.11(b) Part of the sculptors' frieze, the work of John Birnie Philip.

to be the 'soul' of the memorial, and contains 169 life-size, full-length sculptures in a mixture of low- and high-relief. The depictions of earlier figures necessarily were imaginary, although many of the figures were based on materials contained in a collection of artworks and drawings gathered for the purpose of ensuring authentic depictions where this was at all possible.

Two sculptors, Henry Hugh Armstead and John Birnie Philip, were commissioned for this mammoth task. Noting that the monument faces approximately south (Fig. 7.1), Armstead carved the figures on the south and east sides, i.e., the poets' and musicians' frieze and the painters' frieze, respectively (80 figures in total), see Figure 7.11(a), and grouped them by national schools. 'Birnie' carved the figures on the west and north sides, i.e., the sculptors' frieze and architects' frieze (89 named figures, plus two generic figures), see Figure 7.11(b), and arranged them in chronological order. The carving was executed *in situ*, and, although the structural elements of the memorial were completed in four years, it took another four years for the completion of the sculptures and the embellishments, which were finally finished by 1872. Armstead is reported to have spent six years working on the frieze.

Known as the Parnassus frieze, it is named after Mount Parnassus, a sacred mountain in central Greece where Delphi, famous as the ancient sanctuary of the Oracle that was consulted on important decisions throughout the ancient classical world, occupies an impressive site on the south-western slope of the mountain. The fountain of Castalia, a spring near Delphi, is, according to some traditions, the home of the Muses and, as such, Parnassus became known as the home of poetry, music and learning. Another factor that makes the naming of the frieze so appropriate was the existence of the 'Parnassian Movement' which was comprised of a group of young poets writing in mid-19th century France, i.e., Albert's exact contemporaries, who were characterised by their concern for craftsmanship, objectivity and lasting beauty and who emerged as a reaction to the less-disciplined types of Romantic art. The Parnassians strove for exact and faultless workmanship, selecting exotic and classical subjects that they treated with rigidity of form and emotional detachment. Albert, a 'Renaissance man' living in the 19th century, was a Parnassian *par excellence* and the style of the frieze captures this artistic mood precisely.

In discussing the sculptures of the monument, "Study of the monument in its social and architectural context", Bayley (1981) notes that:

> The sculpture is the purpose of the Albert Memorial, all the decorative arts combined with architecture are there to set off either the memorial statue of Prince Albert himself or to elucidate further the various groups of statuary whose job it is to enhance, by the agency of didactic art, the Prince's reputation as a humane bond between the continents on the one hand, and the arts and sciences on the other. . . . The focus is Prince Albert. He is surrounded by sophisticated marmoreal (i.e., of marble) emblems of the continents, the skills of commerce, manufacturing and engineering, the Parnassus of the fine arts from David to Beethoven, Homer to Goethe, Hiram to Charles Barry which Victorian Britain considered to be its legitimate inheritance and the basis upon which its own art flourished.

As noted earlier, Scott's idea was that the memorial architecture and sculpture should, as in the Middle Ages, work hand in hand in selfless harmony. Bayley notes that "This ideology had in part been inspired by Charles Robert Cotterell's study of Wells Cathedral (1865–7) and Scott spoke of uniting sentiments of the Greek and of the Gothic, the perfection of the former being enhanced by the warmth of the latter. In these two sculptural styles, Scott found a 'marvellous consanguinity' and, in bringing the two in contact, he saw what he considered to be the future of art."

The quantity and quality of the sculpture on the Memorial are indeed impressive and at least 12 artists, many of them leading sculptors of the day, were involved in its execution. In addition to the frieze, the eight sculptural groups, the four continents positioned at the four corners of the pyramid of steps that lead up to the monument, and the four 'industrial arts', Agriculture, Manufacturers, Commerce and Engineering, are also carved from Campanella marble, Figures 7.2 and 7.14.

The pedestal on which the statue of Albert sits is supported by the central brick column of the undercroft, Figures 7.2 and 7.3. It is clad in granite and marble, Figure 7.12. Corennie granite is employed for the top of the structure and Ross of Mull granite for the base. (Note that Ross of Mull granite is also the granite used for Stevenson's Skerryvore lighthouse

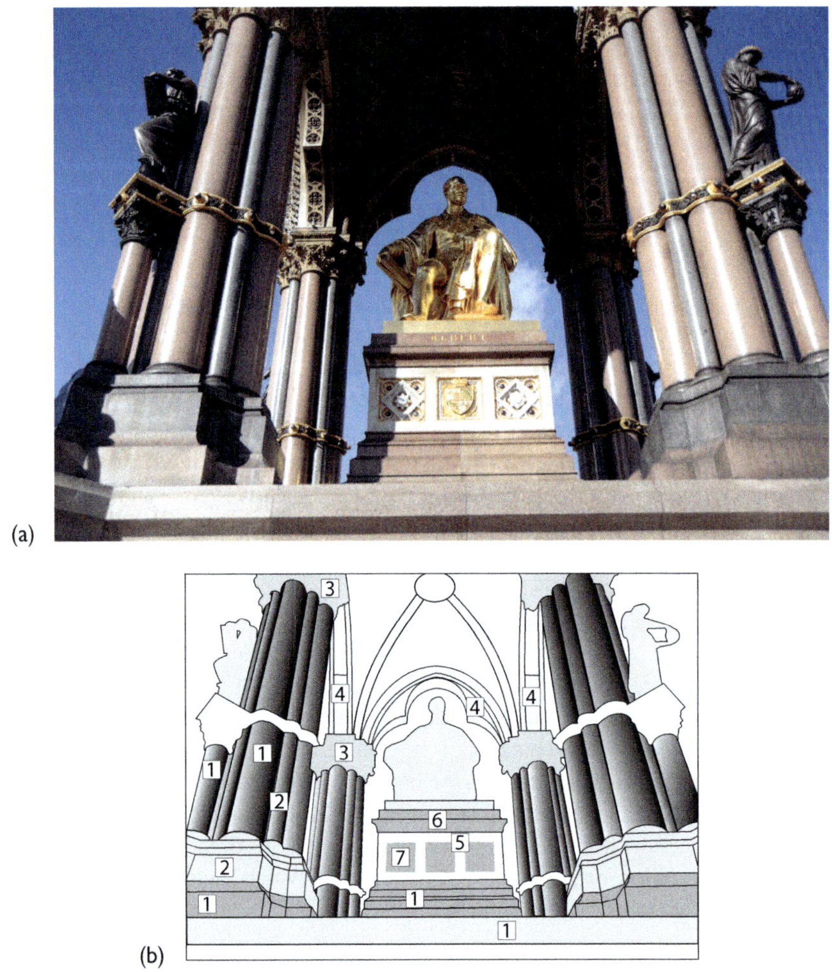

Figure 7.12 (a) The granite and marble clad pedestal supporting the gilded bronze statue of Prince Albert is surrounded by the pillars supporting the canopy. The thicker, pink columns are of Ross of Mull granite and the thinner, grey columns of Castlewellan granite. (b) Rock types used for the pedestal for the statue of Prince Albert and the columns and arches for the canopy. 1. Ross of Mull granite; 2. Castlewellan granite, County Down, Northern Ireland; 3. Darley Dale sandstone, Derbyshire; 4. Portland limestone; 5. Campanella marble; 6. Corennie granite, Aberdeenshire; 7. Chivalric Orders & ceramic coats of arms.

Figure 7.13 The enameled coats of arms of Queen Victoria (left) and Prince Albert (right) by Skidmore are set into the band of Campanella marble that surrounds the pedestal (east-facing side). The pedestal is capped by granite from Corennie in Aberdeenshire and its base is of granite from the Ross of Mull, Argyll and Bute.

highlighted in Section 5.2.2) The Corennie granite comes from Corennie in Aberdeenshire on the Scottish mainland and, like the Ross of Mull granite, was intruded during the late stages of the Caledonian orogeny some 420 million years ago during the Late Silurian or Early Devonian times. Its colour, texture and composition are very similar to the Mull granite and the reason for its use at this location in the Memorial (in preference to its equivalent from the Inner Hebrides) is unclear. The pedestal is encircled with a band of Campanella marble into which enameled coats of arms are set, Figure 7.13. These are by Francis Skidmore, who was responsible for the metalwork structure and decoration of the fleche, see Section 7.2.7.

7.2.6 Canopy (structural and decorative features)

The canopy, Figure 7.14, is supported by four pillars, each consisting of eight granite columns surrounding a central core. The four larger columns of each cluster are made from highly polished, red Ross of Mull granite (0.6 m, 2 ft, in diameter at the base and slightly tapering upwards) and the thinner (less than 0.3 m, 1 ft, in diameter) grey columns from the Castlewellan granite from Ireland. Each cluster of columns is surrounded by an embellished and bejewelled bronze band, Figures 7.14 and 7.15, positioned at one-third of their height. However, the structural implications of the bands, which appear to tie the columns together and therefore enhance their strength and stability, is an illusion. In fact, they hide an inherent weakness in the structure: namely, that the pillars are not monoliths but are made of two separate sections, a feature that facilitated their manufacture (which occurred on site using portable lathes) but which detracts considerably from their mechanical stability. However, as pointed out by Bayley (1981), stability was ensured by other means. The four main

Figure 7.14 The canopy constructed on the podium. The south (sunlit) and east facing sides can be seen together with 'Manufacturers', one of the four 'Industrial Arts' sculptural groups.

pillars, i.e., Ross of Mull columns in each cluster, each weighing 17 tons (~15 tonnes), are dovetailed into a central core and sealed with cement, Figure 7.16a(ii). For greater structural strength, the granite columns were joined one-third of their length from the base, while the core was joined the same distance from the capital, thus giving a sturdy overlap. Copper cramps and dowels Figure 7.16a(iii) maintain a static performance every bit as good as if the multiple columns were in fact monoliths.

The bases of the pillars are made of two parts: a lower section of Ross of Mull granite comprising several blocks and an upper single block of a dark variety of Castlewellan granite, Figure 7.12(a). The work on these pillar bases was highly regarded at the time of construction. The blocks of Castlewellan granite for each base were single stones, each of which, before being finished, weighed about 17 tons (~15 tonnes). The working of each

Figure 7.15 One of the clusters of columns supporting the canopy showing the elaborate Corinthian style capitals sculpted from Darley Dale stone. Note the decorated bronze band concealing the joint between the two sections of the columns. The bronze statue is an allegorical representation of Astronomy; one of the four sciences placed on 'stub' columns in front of the columns of the canopy; see also Figure 7.14 which shows the two sciences, Geometry and Chemistry, respectively, to the left and right of Astronomy.

employed eight men for about 20 weeks and is probably one of the most highly finished and costly pieces of work executed in granite in relatively modern times.

The pillars are capped with 'Corinthian style' capitals (see Fig. 5.6 in Chapter 5 where pillar styles are illustrated), each weighing 13 tons (~12 tonnes). They are sculpted from Darley Dale stone, Figures 7.14 and 7.15, and together support the canopy roof, which consists of four arches above each of which is a triangular gable. These arches and triangles are defined by Portland limestone sculpted by William Brindley, whom Scott describes as "the best carver I have met with." Scott also notes that, "The cornices of the canopy are carved with noble foliage in high and bold relief much of which is gilded and embellished with semi-precious stones and coloured glass", Figures 7.15–7.18. The individual elements of the high relief foliage are termed 'crockets'. These hook-shaped decorative features are common in Gothic architecture and are so called because of their resemblance to a bishop's

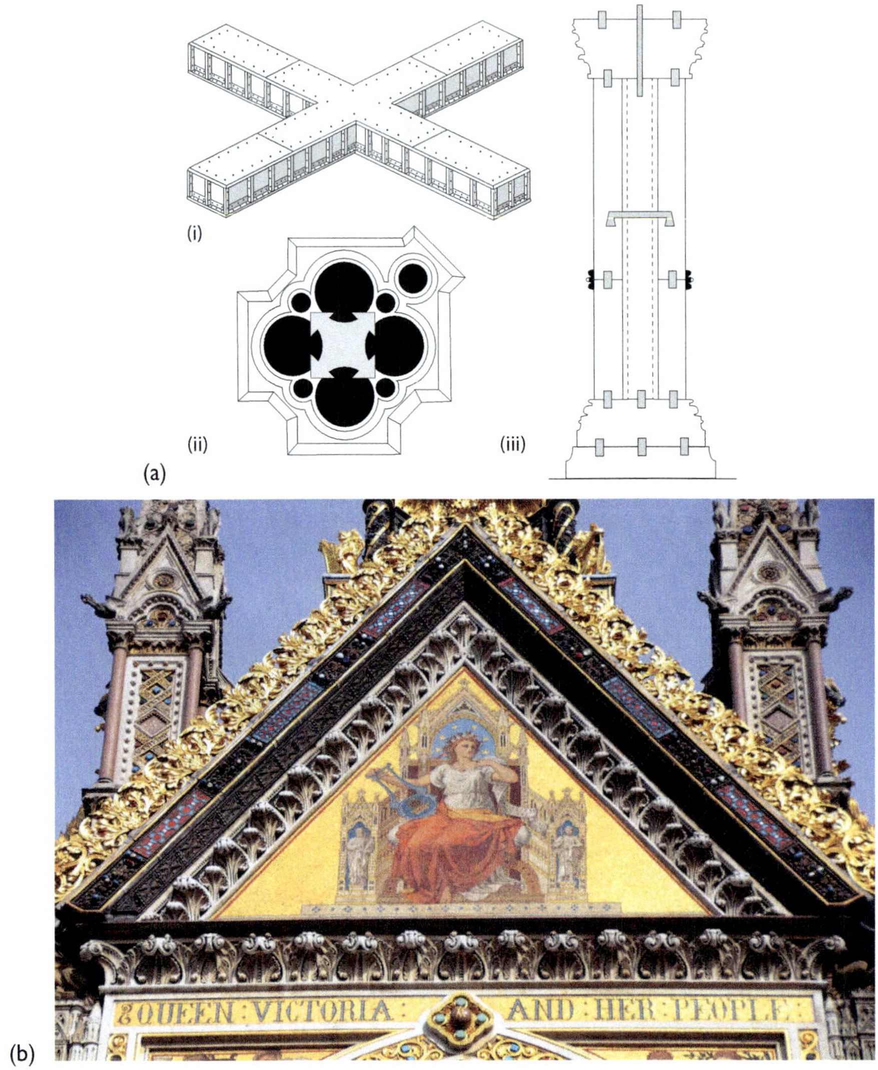

Figure 7.16 (a) (i) Cross box girder on which the fleche rests. (ii) Cross-section through one of the four cluster pillars that support the canopy. Note the dovetailed linking of the four main columns to the central core. (iii) A vertical section through one of the four cluster pillars showing the copper clamps that stabilise the structure. (b) A mosaic representation of Music on the south facing gable. Panel (a): (Modified from an engraving in *The National Memorial to His Royal Highness the Prince Consort*, 1873, Kensington Central Library).

crosier. The gables are further enriched by ornate gilt bronze 'crestings' along the gable edges and roof ridges of the canopy, Figures 7.17 and 7.19.

The mosaics which fill the Tympana (the semi-circular or triangular decorative wall surface bounded by a lintel and arch), Figure 7.16(b) and spandrels (the space between the

Figure 7.17 Portland limestone used in the framing of the triangular gable and spandrels (i.e., the triangular spaces between the arch and the surrounding right-angled framework) of the south facing front of the canopy.

arch and the rectangular enclosure), Figure 7.18, were designed by John Richard Clayton. They depict figures representing Poetry and Music, (south facing), Painting (east facing), Architecture (north facing) and Sculpture, (west facing) reflecting the four arts celebrated on the equivalent four faces of the Frieze of Parnassus which surrounds the podium below. The tesserae (individual pieces) of the mosaics are not ceramic but are made of Venetian (Murano) glass provided by Antonio Salviati, who had recently revived the ancient Venetian arts of glass and mosaics. Other materials used to embellish the monument include enamel, polished stone, agate, onyx, jasper, cornelian and crystal (i.e., quartz).

Each allegorical figure sits on an ornate gothic throne which is flanked by two smaller figures representing two great historic artists in that field. Music and Poetry (Poesis) has David and Homer; she also carries a little scroll which lists Shakespeare, Moliere, Milton, Goethe, and, oddly, Homer again; and Sculpture (Sculptura) has Michelangelo and Phidias—the Greek sculptor, painter and architect whose statue of Zeus at Olympia was one of the Seven Wonders of the Ancient World. Painting (Pictura) has Apelles of Cos and Raphael; Architecture (who holds a design of the Albert Memorial itself) has Solomon (who built the Temple of Jerusalem) and Ictinus (credited as being one of the architects of the Parthenon).

Eight bronze statues representing the practical arts and sciences decorate the canopy. They were originally intended to be in marble and are the work of Armstead and Philip. Four

Figure 7.18 Sculpted Portland limestone gilded and embellished with semi-precious stones and glass outlining one of the spandrels depicting poetry on the south-facing side (i.e., the front) of the canopy.

of these are placed on 'stub' columns in front of the pillars, Figures 7.12, 7.14 and 7.15. They are Astronomy (SE) and Chemistry (NE) by Armstead, and Geology (NW) and Geometry (SW) by Philip. The remaining four statues, Figures 7.14, 7.17 and 7.18, are positioned in niches in the canopy directly above the capitals and adjacent to the spandrels. They are Rhetoric (SE) and Medicine (NE) by Armstead, and Philosophy (NW) and Physiology (SW) by Philip.

Armstead was sufficiently satisfied with his Astronomy, Figure 7.15, that he exhibited it independently at the Royal Academy in 1868. In addition, his Chemistry was much admired and the Pall Mall gazette (14 July 1869) found it "simple and vigorous looking" and that "there is in the whole design a frank hardihood and a scorn for prettiness." However, in the context of the present work, it is Philip's Geology situated on the NW angle of the canopy that deserves our attention. The figure has a hammer or pickaxe in her right hand and a

Figure 7.19 One of the four Portland limestone pinnacles built directly above the canopy pillars. The photograph also shows the gilded bronze cresting along the edge of the gable and the elaborate lead-working, gilding and inlaying of vitrified enamel along the gable edge and on the fleche below the upper shrine which houses the eight Virtues.

partly exhausted globe in her left, which (with remarkable prescience) illustrates the depredations which the search for precious minerals was to make on the Earth's surface. At her feet, the same metallic ores and antediluvian remains characterise further the business of the geologist.

The roof of the canopy is clad in lead and decorated with detailed lead work. Four pinnacles positioned directly above the four supporting pillars rise from the roof, Figures 7.2, 7.17 and 7.19. They are carved from Portland limestone and consist of a rectangular base beneath a gabled roof from which four small rectangular pinnacles spring, Figure 7.19, mirroring the geometry of the main canopy of the memorial. Each small pinnacle is surmounted by a lion sitting on its haunches (i.e., a lion 'sejant'). The four main pinnacles are capped by square-based pyramids with intricately carved and bejewelled faces, and crowned with a gilded finial. The rectangular base of each pinnacle has a red granite pillar at each corner and a panel is carved into each face and decorated with embossed lattice work. Within the central lattice space the entwined initials V and A have been carved and the other spaces are occupied by coloured glass and polished stones. Albert's motto "*Treu und Fest*" (Faithful and Fixed) is engraved along the struts of the lattice, Figure 7.19. Gargoyles extend from the four corners of the canopy roof, Figures 7.14 and 7.17. These were designed by Scott and carved from blocks of Darley Dale stone by William Brindley.

7.2.7 The fleche[1] (steel structure but purely decorative)

The monument can be divided neatly into two distinct parts in terms of the materials from which it is constructed. The structure is predominantly of stone and bricks from the foundation to the Canopy but the 'fleche' is predominantly of metal that rises majestically from the canopy to the full 55 m (180 ft) height of the embellished bronze cross which surmounts the structure. Scott entrusted all the metalwork of the Memorial, both structural and decorative, to Francis Skidmore, a renowned and extremely skilled artist who had previously been commissioned by Scott to make the iron and brass choir-screens for Exeter College chapel in Oxford. He had also made elaborate screens for the Worcester and Salisbury cathedrals.

The fleche has an internal framework of iron and the whole weighs 210 imperial tons (213 tonnes). This rests on a cruciform box girder, Figure 7.16 a(i), itself weighing 21 tonnes, whose four ends rest on the four pillars of the canopy. A vertical box girder of bolted cast iron provides a central core to the fleche, which supports the iron frame to the canopy roof. At a yet higher level where the fleche tapers, the core becomes a cluster of columns, and finally a single shaft which supports the orb and cross surmounting the monument (Fig. 7.20).

The fleche is clad in lead, at least 0.25 inches (~6 mm) thick, needed to protect the ironwork from 'atmospheric influences' using a large quantity of bronze screws. Scott had suggested the use of ornamental lead work on the canopy roof, and Skidmore relished the opportunity of "elevating a tame and generally cumbrous material into one of adornment and beauty, by the use of geometric patterns and vitrified enamel inlays", Figures 7.19 and 7.21.

It will be recalled that Scott intended the memorial to be a full-sized realisation of the 'Jeweller's architecture' of the medieval shrine makers and in his *Recollections* he notes that:

> The parts (of the memorial) in which I had it in my power most literally to carry out this thought were naturally the roof with its gables, and the fleche. These are almost an absolute translation to the full-size of the jeweller's small-scale model. It is true that the structure of the gables with their flanking pinnacles is of stone, but the filling in of the former is of enamel mosaic, the real-size counterpart of the cloisonné of the shrines, while all the carved work of both is gilded, and is thus the counter-part of the chased silver-gilt foliage of shrine-work. All above this level [i.e., the fleche] being of metal, is literally identical in all but scale, with its miniature prototypes. It is simply the same thing translated from the model into reality.

As the fleche rises from the canopy the geometry of the vertical box girder is apparent, Figures 7.20–7.22. The lead cladding is patterned and sculpted and has decorative panels containing areas of finely wrought ornamental lead work embellished with enamel

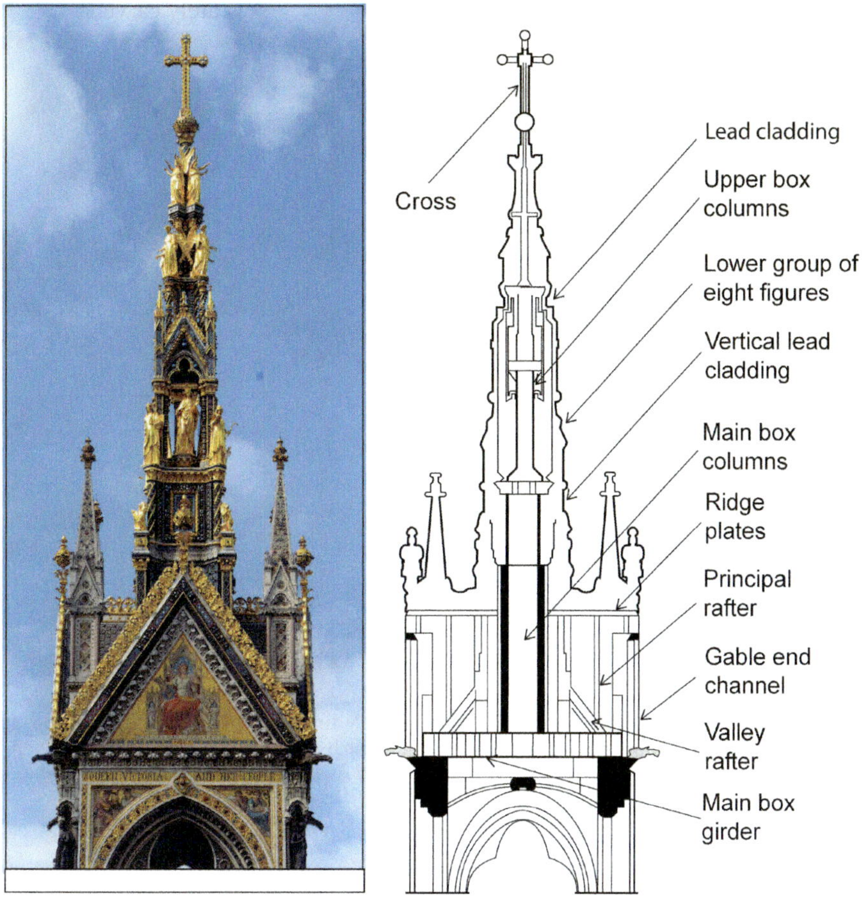

Figure 7.20 Sketch and photograph showing the main structural features of the metal fleche (spire).

Figure 7.21 Examples of lead work. A lower panel of the fleche, left. Right an upper panel of the fleche directly below the small, upper shrine, see Figure 7.19.

Figure 7.22 The lower portion of the fleche showing clearly the geometry of the vertical box girder, see Figure 7.20. The structure is clad in lead which is patterned, sculpted and embellished with decorative panels above which is the small shrine containing the eight gilded virtues.

inlays and large coloured glass bosses set in bezels, Figure 7.21(a) and (b). At this level, the fleche is flanked by four gilded bronze heraldic lions, each carrying a scroll on which Prince Albert's motto "*Treu und Fest*" (Faithful and Fixed) is inscribed. These lions, together with the upper panels of the fleche, Figure 7.21(b), act as a pedestal to a small shrine, Figure 7.19. Its design is based on the shrine of Edward I in Westminster Abbey and reflects the form of the memorial itself, styled on the twelve Gothic Eleanor crosses which marked the resting places of Edward's Queen, Eleanor of Castile, when in 1290 her body was returned from her deathbed in Harby, near Lincoln, to Westminster Abbey in London. The shrine houses four Christian and four Moral or Cardinal virtues, modelled by James Redfern (1838–76).

In the same way as the corners of the main podium of the memorial are extended to provide supports for the sculptural groups representing the Industrial Arts, so the corners of the pedestal for the tabernacle containing the eight virtues are extended to support the four moral virtues, Justice, Fortitude, Prudence and Temperance. These are more prominent, being set at the corners of the shrine and face NW, SW, SE and NE respectively, Figures 7.19 and 7.20 . Less prominently displayed are the four Christian virtues of Faith, Hope, Charity and Humility; these occupy the four niches of the shrine and face north, west, south and east, respectively. All eight statues are gilded and are thought to be of bronze.

Two tiers of gilded angels, modelled by Philip, adorn the upper stages of the fleche. The lower four have their arms inclined downwards in an attitude suggestive of the resignation of worldly honours. In contrast, the four angels of the upper tier have their arms uplifted in an act of worship. The fleche is topped by a large bejewelled and gilded bronze cross on the top of an orb, Figure 7.20. The original cross was shot down by anti-aircraft fire during the Second World War but was eventually replaced in 1955. During the inspection of the monument in 1985, prior to its major renovation, further damage was discovered to one of the Virtues. *The Guardian* newspaper reported on 4 July 1985 that, "An architectural survey by the government has revealed one of Britain's oddest wartime own-goals: a direct hit on the Albert Memorial by an anti-aircraft gun. The statue of Humility was hit in the shoulder and a pigeon had built its nest in the three-inch wide hole through the bronze. The shell is thought to have come from an ack-ack battery sited on the nearby carriage drive through Hyde Park. The trajectory suggests that an over-enthusiastic gunner tracked a bomber too far to the west."

During the planning stages of the Memorial, and as a contribution towards the metal required for the structure, the government had promised to donate 37 redundant cannons captured in the Crimean War and stored at the Woolwich arsenal. These were of bronze (an alloy of about 90% copper and 10% tin) or gun metal. Some of these were melted down and used for statues in the fleche and for the gilded crestings that cap the gables, capitals and cornices. The cannons are also thought to be the source of bronze for Foley's monumental statue of Albert, the centrepiece of the memorial, Figure 7.12(a). The model Foley had created, which had been placed on the memorial prior to casting, was moved to the studio and cut into approximately 1500 pieces, each of which was moulded and cast, then welded and screwed together before being finished and gilded.

As Sheppard (1975) notes, Scott was delighted with Skidmore's work on the fleche, commenting that:

> I have been enabled to realise most exactly the ideal I had in view. . . . With copper and lead-covered iron, Skidmore reproduced, in noble workmanship, and to a noble scale, the repoussé work, the chased and beaten foliage, the filigree, the gem-settings, and the matrices for enamels of the mediaeval gold- and silver-smiths. No nobler work in metal for architectural purposes has, so far as I know, been produced in our own, or, probably, considering its scale and extent, in any other age; nor do I think that any man living but Mr. Skidmore could have produced such a work.

* * * * *

Like it or loathe it, the Albert Memorial is the greatest public monument ever to be erected in Great Britain. It is the masterpiece of Sir George Scott and is one of the greatest works of Gothic revival in the world. The Memorial in Kensington Gardens in London can be viewed at any time, and the authors strongly recommend readers to make a visit.

7.2.8 Albertopolis

The term 'Albertopolis' refers to the area shown in Figure 7.23. Prince Albert, who was a promoter of the very successful Great Exhibition of 1851, also had the vision that the area in the photograph should be dedicated to the arts and sciences and the achievements of Victorian Britain. Note that we have already highlighted the Natural History Museum in Figures 3.96–3.98 when discussing faience (glazed terracotta), and this Museum on Cromwell Road can be seen in the foreground of Figure 7.23. The street on the left is Queen's Gate and the

Figure 7.23 The South Kensington buildings—from the Natural History Museum in the foreground through to the Albert Memorial just beyond the dome of the Albert Hall near the top of the photograph.

street on the right is Exhibition Road. Travelling from the Natural History Museum along the vertical centreline of Figure 7.23, there is a white tower, the Queen's Tower, which is the last remnant of the original Imperial Institute, the forerunner of Imperial College London, now having its main entrance located on Exhibition Road. Moving further up the centreline of Figure 7.23 is the circular Albert Hall, and beyond that the Albert Memorial can just be seen located in Kensington Gardens adjacent to Hyde Park. Also included in this area are the Royal College of Art, the Science Museum and the Victoria and Albert Museum, which is just outside the lower right part of the photograph.

7.3 DURHAM CATHEDRAL

Durham Cathedral, Figure 7.24, in the north of England was originally a fortified Benedictine cathedral-monastery and is an exciting Romanesque structure from three main points of view:

- the location, majesty and beauty of the building;
- the construction and architecture of the cathedral, starting in the 11th century and which includes the first use of the pointed 'Gothic arch' in Britain; and
- the type of sandstone used for the cathedral's construction, together with the associated specific nature of the building stone weathering.

We address these three aspects in the following sections.

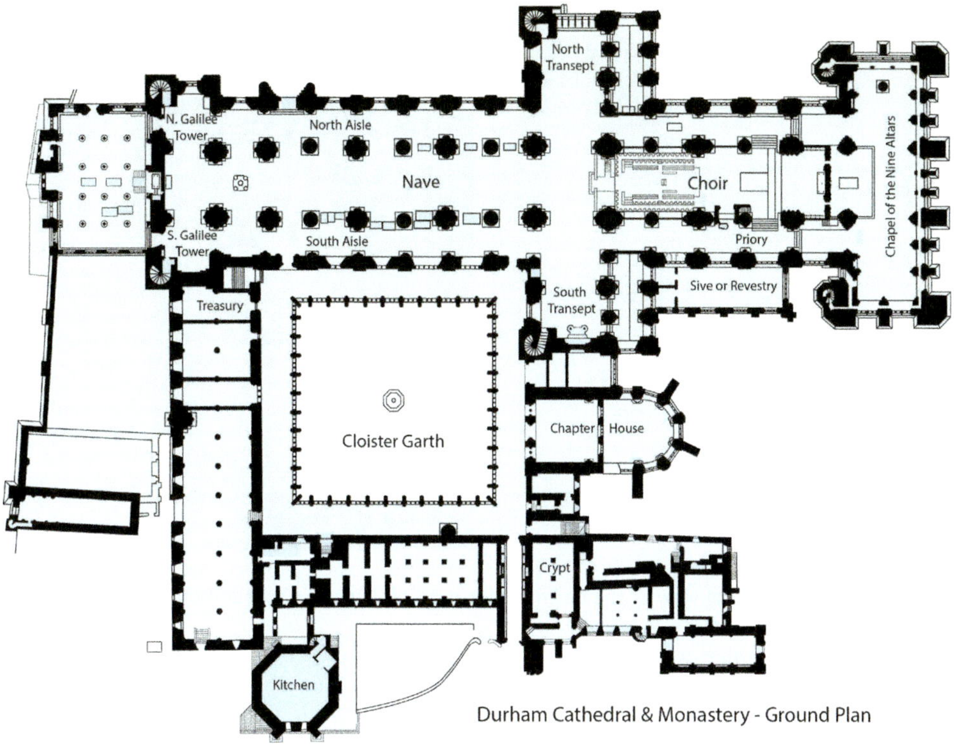

Figure 7.24 "Durham Cathedral is *par excellence* the greatest of all our Norman churches" (Cook, 1951). Its length is approximately 150 m.

7.3.1 Cathedral location

The topography on which Durham Cathedral is situated is a consequence of an incised meander of the river Wear in northern England. This formed as a result of the regional uplift of the land in response to the removal of the ice load at the end of the last glacial period—approximately 12,000 years ago. The rate of uplift was slow and the meandering river was able to maintain its course by the process of erosion as the uplift occurred. The elevated rock promontory on which the cathedral and the nearby castle are situated, Figure 7.25, corresponds to one incised meander loop. The strata comprising the promontory are Upper Carboniferous in age and belong to the Middle Coal Measures formation, which is made up of mainly sandstones with interbedded mudstones and coal seams. These sandstones were called 'posts' by local quarrymen and miners and the stone for Durham Cathedral, Lower Main Post sandstone of the Middle Coal Measures, was sourced mainly from Kepier Quarry, less than two miles to the north-east. The location of the quarries is in the vicinity of Durham town itself, although they have long been abandoned.

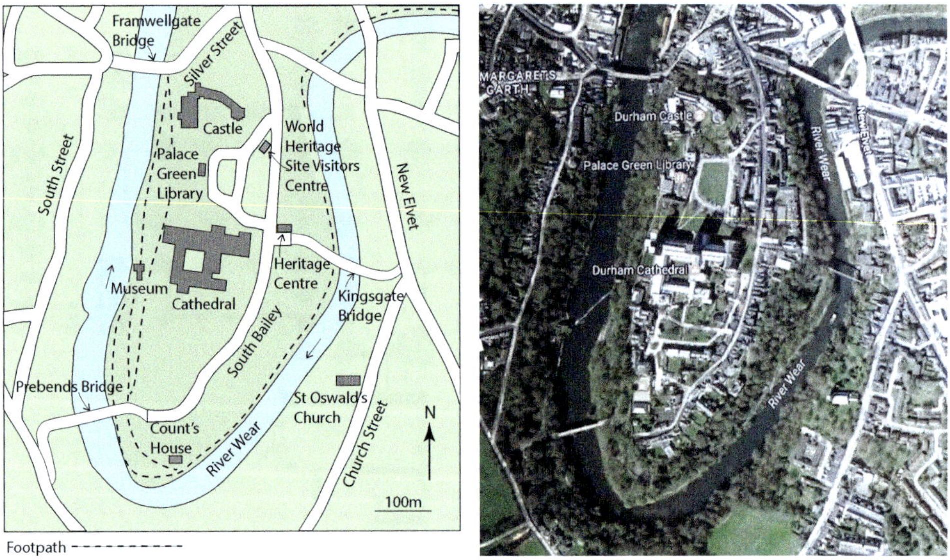

Figure 7.25 Map and aerial view of the defensive location of Durham Cathedral on a rocky peninsula above the river Wear, England.

The setting of Durham Cathedral is eloquently described by James F. Hunnewell in his book *England's Chronicle in Stone* (1886) where he notes that:

> Durham presents the noblest view of its kind in England. From high land the ground slopes steeply to the river Wear, the valley of which is filled with the houses of the city. Beyond, and over them, precipitously rise high, rocky or tree-clad banks, and, along their extended crest, the vast and wonderfully picturesque walls of the embattled castle, and the tower-crowned cathedral; while behind these grey and venerable monuments, and far into the hazy distance of the background, stretch the verdant Northumbrian hills. In all the world, there are few such majestic groups of mediaeval works so nobly placed. . . . Long and lofty in form, austere and massive in its Norman style, crowned by three great towers, the venerable church stands in calm strength and majesty on its high base of rock, above a zone of fresh and graceful tree-tops, that with their bright waving green relieve its sombre tints, and show by their marked contrast its bold and immovable grand walls.

This description is still valid today: the twin towers at the west front of the Cathedral are shown in Figure 7.26, and the adjoining castle is shown in Figure 7.27.

In addition to its setting and architecture, the cathedral contains a superb medieval Benedictine Library, having manuscripts dating back to the 6th century, plus three Magna Carta versions (1216, 1225 and 1300). The Magna Carta was issued by King John of England in the 13th century, establishing for the first time the principle that everybody, including the king, was subject to the law.

Figure 7.26 The twin towers of the West Front of Durham Cathedral (65,000 tonnes of sandstone used) viewed from the river Wear.

Figure 7.27 Durham Castle viewed from the river Wear.

7.3.2 Construction and architecture of Durham Cathedral

The dimensions of this early Norman masterpiece, the building of which began in the late 11th century, are as follows. Internal dimensions: overall length 143 m (469 ft), length of nave 60 m (198 ft), width of nave and aisles 25 m (81 ft), Figure 7.24. Height of central tower 66 m (218 ft), height of western towers 44 m (144 ft), Figure 7.26.

The cathedral, constructed to house the shrine of St. Cuthbert and where the Venerable Bede is buried, is built of Coal Measures sandstone quarried at Kepier, which is a mile east of Durham, and was ferried upstream on the river Wear to the cathedral site. On 2 August 1093, the foundation of the cathedral was laid, after which construction proceeded rapidly from east to west, this direction being favoured so that the section with the altar, and hence a place of worship, would be created early in the process. A large part of the cathedral was constructed in the next six years and then, between 1099 and 1128, the nave and the western towers were built, with construction being completed in 1133. Needless to say, in the succeeding centuries, many structural alterations have been made. The present day interior of the cathedral is shown in Figures 7.28 and 7.29.

Figure 7.28 Interior of Durham Cathedral. Note the presence of the large incised pillars and both the semi-circular and pointed roof arches. See Figure 5.61 and associated text in Section 5.3.5 for an explanation of the geometry of the roof vaulting.

Figure 7.29 The interior of Durham Cathedral, noting the incised pillars.

However, the construction of the cathedral, with its homogeneous Norman-Romanesque style and incomparable unity and majesty, was not an isolated building project. Johnson (1980) notes that:

> In 1088, the Bishop of Durham, William of St. Calais, quarrelled with the new king, Rufus, and was sent into exile in France for alleged treason. He spent three years there, and had the opportunity to examine some of the latest abbey-churches and cathedrals in Normandy and the Ile de France [Parisian region]. There he seems to have conceived the idea of a gigantic new cathedral, made possible by the concentration of wealth and resources in the Durham Palatinate—a status given to Durham by William the Conqueror and which extended from the river Tyne to the river Tees. On his return in 1093 he demolished the Saxon church, and the foundation stone of the new cathedral was laid the next year.

Johnson (1980) also notes that, during the years 1093–1133, when Durham Cathedral was being built, more large-scale buildings were being erected in England than anywhere else in the world. The cathedral, actually a cathedral–priory, and containing the relics of St.

Cuthbert, was served by monks of the Benedictine order. Services were held regularly from the time of construction until the suppression of the monasteries by Henry VIII.

In particular, Durham Cathedral, architecturally the best Norman structure in Britain, is noted for its massive internal incised pillars, Figure 7.28–7.30, and the use of pointed Gothic stone arches supporting the stone roof, Figure 7.28. This type of arch is more efficient than the semi-circular Roman–Norman arches, see Figure 7.29, because a pointed arch is taller, enabling the windows to be higher. In fact, Durham Cathedral has the country's first structural pointed arch and stone roof, marking a turning point in the history of British architecture, see Section 5.3.5.

The cathedral is a classic example of Anglo-Norman/Romanesque architecture which, with its size and solidity, must have been a dramatic successor to the less significant earlier Saxon stone buildings—and indeed the structure is dramatic to this day. The main characteristic of such Norman stone structures is architectural dependence on the square and circle and their 'massiveness', both physically and visually, and especially as manifested by the internal patterned circular stone pillars 8 m (27 ft) high and 2 m (7 ft) in diameter, noting that in many cathedrals these pillars, having a stone exterior, are filled with rubble. The heavy piers/pillars alternate as clustered piers and large round pillars, some decorated with the typical Norman chevrons and chequerboard patterns, as seen in Figure 7.29.

Close-ups of the decorative pillar styles are shown in Figure 7.30, with the captions indicating how many separate types of building stone incisions would be required for each overall decorative style. The pillars are 9 m high and, with their different incision patterns (diaper, chevron, flute and spiral), have a majestic presence. Masons prepared the standardised building stones in the winter and emplaced them in the summer (when the mortar set more easily), the interiors of the pillars being filled with rubble. Note the error in the pillar pattern in Figure 7.30(c), possibly as a purposeful demonstration that humanity cannot achieve perfection.

Decorative composite pillars are also present, e.g., Figure 7.31, which illustrates the use of Frosterley 'marble'. This is obtained from the nearby village of Frosterley and is used for fonts, floors and columns in many churches and other buildings. Like the Purbeck 'marble' shown in Figure 3.27, this marble is also a limestone. It will be recalled that geologists use the term 'marble' to refer to metamorphosed limestone; however, stonemasons use the term more broadly to encompass some unmetamorphosed limestones. The internal structures of the Carboniferous fossil corals, *Dibunophyllum bipartitum*, which have curved horn-shaped skeletons and a circular cross-section, are clearly visible in the pillars and in the close-up images in Figures 7.31(b) and 7.32. If this rock had been metamorphosed, the calcite would have been recrystallised and these fossils would have been destroyed.

Figure 7.33 illustrates why pointed arches were such an innovation during construction of the cathedral roof. In order to make the roof higher and thus enable the clerestory windows (i.e., the highest windows) to be larger, hence letting in more light, semi-circular Roman type arches were built, not from directly opposing pillars but diagonally across the nave roof. An example is marked by the white A in Figure 7.33. This meant that arches from directly opposing pillars would be too low if they had a semi-circular shape—and so the pointed Gothic arch, marked with the white B in Figure 7.33, came into being. This was the first time that the pointed arch was used in British architecture; see Section 5.3.5 on roof vaults.

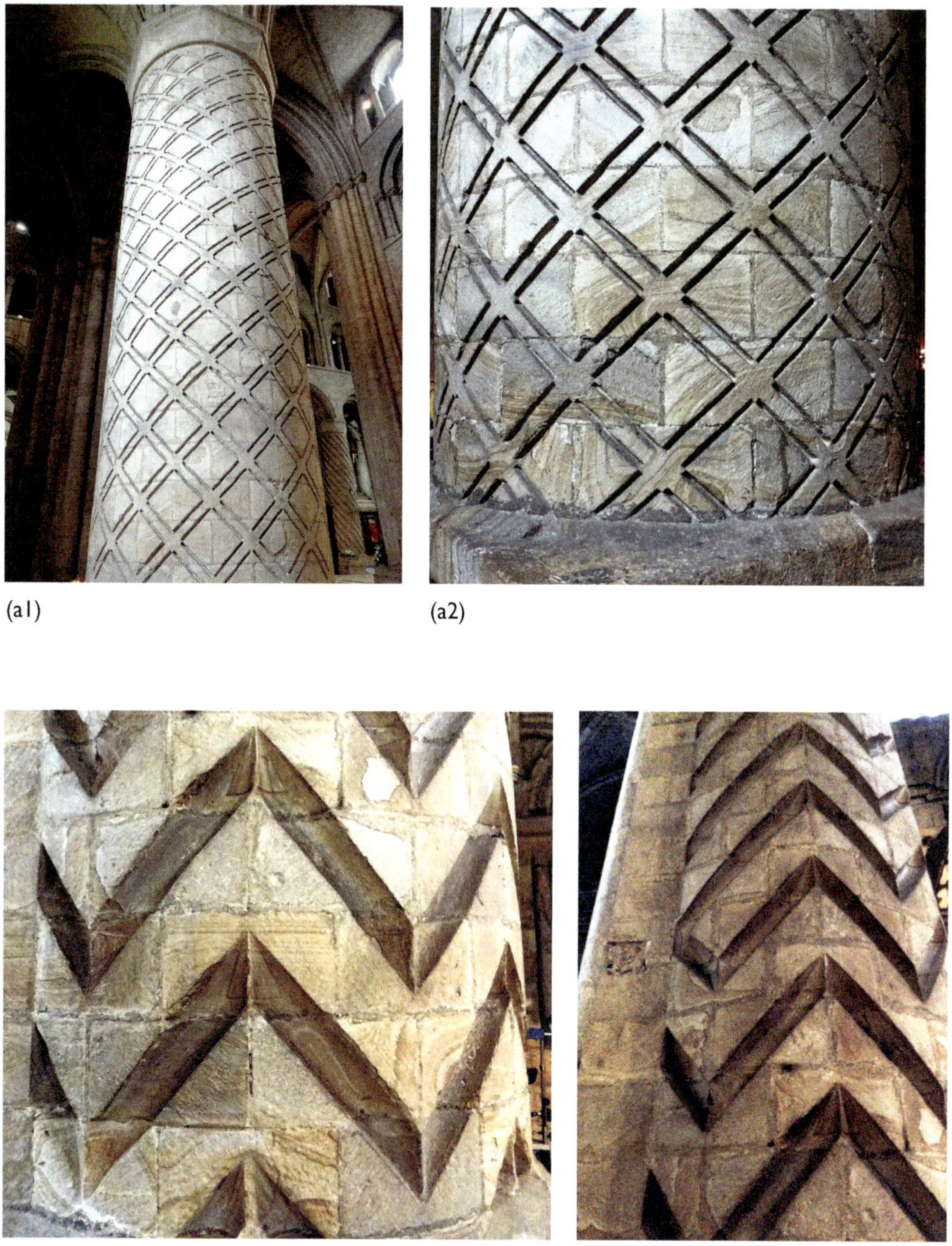

(a1) (a2) (b) (c)

Figure 7.30 Styles of pillar decoration in Durham Cathedral. In (a1) and (a2) only one pattern of building stone incision type is required. For (b) and (c) three types of building stone incision are required. Note in (c) there is an error in the building stone pattern.

(a) (b)

Figure 7.31 (a) Frosterley 'marble' pillars in Durham Cathedral. (b) Closer view of the Frosterley marble with its Carboniferous *Dibunophyllum bipartitum* fossil corals.

Figure 7.32 Frosterley 'marble' containing fossil corals from the Carboniferous period.

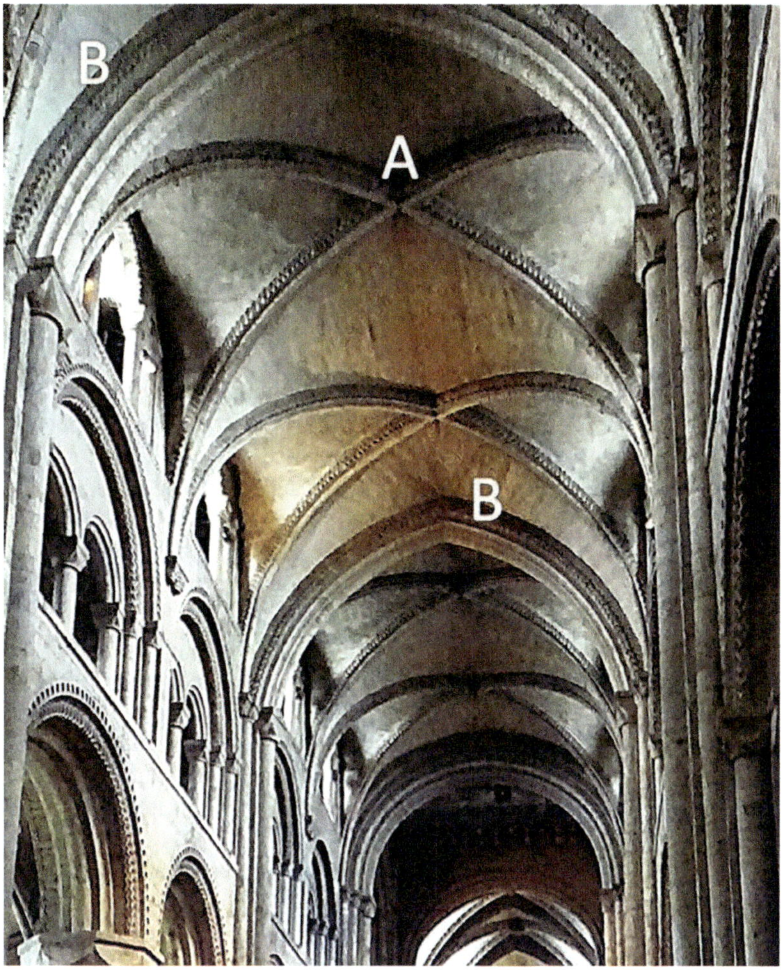

Figure 7.33 Illustrating why pointed roof arches are required in this form of roof geometry. Detail from Figure 7.28 showing A: high semi-circular roof arches from *non-opposing* vertical pillars at the side of the nave, necessitating B: pointed roof arches (i.e., Gothic) for the thicker arches between *directly opposing* pillars.

7.3.3 Weathering of the Durham Cathedral sandstone: a case study

Durham Cathedral provides spectacular examples of sandstone weathering. Attewell and Taylor (1990) have written about "Time-dependent atmospheric degradation of building stone in a polluting environment" with specific reference to Durham Cathedral. They note that the foundations and the main structure of the cathedral were built using the Lower Main Post sandstone of the Middle Coal Measures and explain that the weathered facing stone of the cathedral was undergoing a programme of progressive replacement, i.e., in

1990 at the time when their paper was published. In this context, Attewell and Taylor examined the durability characteristics of the original stone and the factors responsible for the atmospheric weathering which has taken place over almost 900 years,[2] with the last 200 years being particularly significant because of the industrialisation in the north of England. An additional problem was that the Cathedral had no downpipes for draining rainwater.

The weathering has taken the form of deep pitting and exfoliation (spalling) of the exposed faces. Such exfoliation can occur grain by grain or the stone can peel off in sheets. Water is particularly significant in causing building stone degradation, especially in locations where the temperature can cause freezing of the water within the stone—thus expanding and splitting the stone. Attewell and Taylor also note that expansion of iron minerals by oxidation in water can disrupt the stone fabric. The type of pitting observed in sandstones also occurs in limestones: Rodriguez-Navarro *et al.* (1999), in a paper on the origins of honeycomb weathering in limestone, mention that the phenomenon is still not well understood but that heterogeneous wind flow over a stone surface is important in the development of the pattern. Wind also promotes evaporative salt growth between grains on a stone surface, see the following paragraph, resulting in the development of small, distributed cavities.

In a related but later study, Simple's MSc thesis (2010) describes a detailed study of "Alveolar Erosion and its Conservation Recommendations for the Sandstone Masonry at Durham Castle", the castle being adjacent to Durham Cathedral and constructed with the same stone. The word 'alveolar' is an adjective indicating 'like a small cavity, pit, or hollow'. Simple's test results indicate that the building stone is a clay-rich arenite (a sedimentary clastic siliceous rock with less than 15% matrix) containing iron oxide resulting from breakdown of the clay and that, "the stone's deterioration tendencies are primarily a result of the geochemical nature of the sandstone and the weathering is then enabled by repeated contact with moisture and wind and accelerated by the presence of salts."

The salt is gypsum ($CaSO_4.2H_2O$), which is not a primary constituent of the building stone. It is formed when sulphuric acid is generated as the sulphur dioxide in industrially polluted air reacts with water and attacks the calcite within the rock. Durham Castle is shown in Figure 7.34 and an example of its weathering in Figure 7.35.

In the overview paper by Turkington and Paradise (2005) titled "Sandstone Weathering: A Century of Research and Innovation", the authors illustrate the processes of 'building stone retreat' via the processes in Figure 7.36 and the range of spatial scales across which regular pits may be observed on sandstone surfaces in general in Figure 7.37.

7.3.4 Liesegang rings

Liesegang rings are an example of a self-organising, oscillating, geochemical reaction.

The Royal Society of Chemistry, UK

When aged and/or weathered, as is the case at Durham Cathedral, the microstructure of the sandstone, the bedding fabric and the 'liesegang rings' become more apparent. These liesegang rings cause the brown, yellow and white banding in the Figure 7.38 images. The banding in the upper central stone in Figure 7.38(b) is an excellent example of these rings,

Figure 7.34 The main entrance to Durham Castle.

Figure 7.35 Sandstone weathering near the parapet of Durham Castle.

which are not developed when the sandstone is initially formed, but later as a result of the movement of chemical solutions through the sandstone.

In fact, the occurrence of these liesegang rings in sandstone is widespread, not least in the ubiquitous Yorkstone sandstone city pavements (a fine-grained Carboniferous sandstone) found throughout Great Britain, see Figure 3.53. We have also earlier shown examples of the rings in Chapter 3 when discussing the building stones of the 1 Poultry building in London, see Figures 3.48–50 and 52. Whilst the weathering mechanisms described in the previous parts of this section and summarised in Figures 7.36 and 7.37 do operate on the Durham

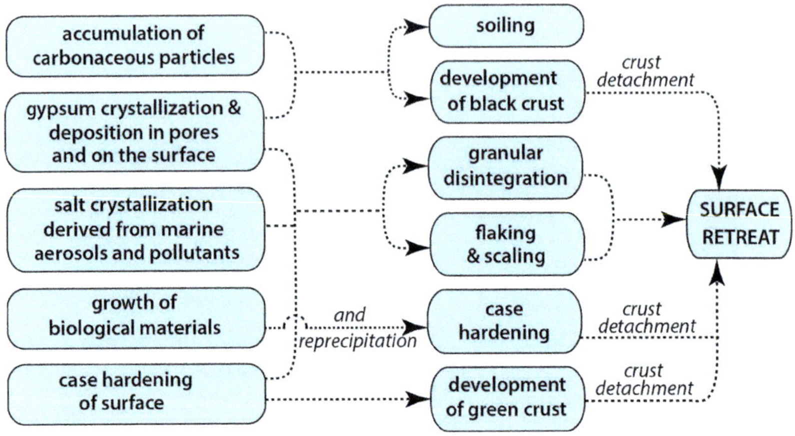

Figure 7.36 Stone retreat processes. (After Turkington A. V. and Paradise, T. R., (2005). Sandstone weathering: a century of research and innovation. *Geomorphology*, 67, 229–253)

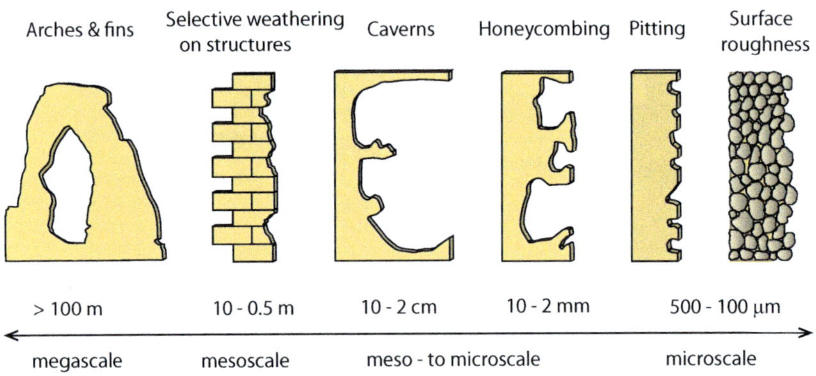

Figure 7.37 The range of spatial scales across which regular pits may be observed on weathered sandstone surfaces. (After Turkington A. V. and Paradise, T. R. (2005). Sandstone weathering: a century of research and innovation. *Geomorphology*, 67, 229–253)

Cathedral sandstone, we concentrate now on these rings because they have a dominant effect on the weathered surface appearance of the building stones.

The patterns of rings or bands are named after Dr Raphael Liesegang, the first scientist to make a systematic study of the phenomenon in 1924. The Royal Society of Chemistry explains that, although there is no universally accepted explanation of the rings, they are an example of a 'self-organising, oscillating, geochemical reaction' which takes the form of a travelling wave and can give rise to complex patterns—as indeed is evident in Figure 7.38. Thus, the liesegang rings are rhythmical precipitation patterns of iron oxides and their formation results in there being two elements that define the microstructure of the rock: the first is the primary fabric generated during sedimentation which is made up of bedding and cross-bedding layering; and the second is the later banded appearance of the fabric, dominated by

(a) (b)

Figure 7.38 The 'liesegang rings' phenomenon in the sandstone building stones of Durham Cathedral.

the liesegang rings, which is the result of variations in the concentrations of iron oxides in the fluids that moved through the rock over geological time. These two fabrics may be concordant, i.e., when the liesegang banding is parallel to the bedding, or discordant when it is not.

Although the building stones in Figure 7.39(a) and (b) appear to be weathered in different ways, in fact all the images are manifestations of the same weathering effect; however, the weathering has different characteristics in the separate building stones as a direct function of the orientation of the bedding and liesegang ring boundaries with respect to the exposed face. The nature of the weathering of some of the individual building stones (indicated by the yellow letters in Fig. 7.39(a) and (b)) is noted as follows.

A: The bedding and liesegang fabrics are concordant in this block and, because these planes are sub-parallel to the block face, flaking has occurred.

B: Mineral veins parallel and sub-parallel to the bedding in the sandstone have dominated the weathering geometry. They now form prominent ridges, which are less susceptible to weathering and stand proud of the intervening layer of sandstone.

C: In this block, the bedding and liesegang fabrics are discordant. The bedding is parallel to the base of the block and the liesegang fabric parallel to the block face, resulting in spalling back to a liesegang surface.

D: The original bedding of the sandstone is apparent in this block: the lower half of the block has horizontal planar bedding, whilst the upper portion exhibits cross-bedding and a greater vulnerability to 'spot weathering', i.e., pitting, Figure 7.37. This block is the 'right way up', with reference to its original bedding orientation, see Section 3.5.3.

E: In contrast, in this block the bedding is steeply inclined and pitting can be seen along the bedding.

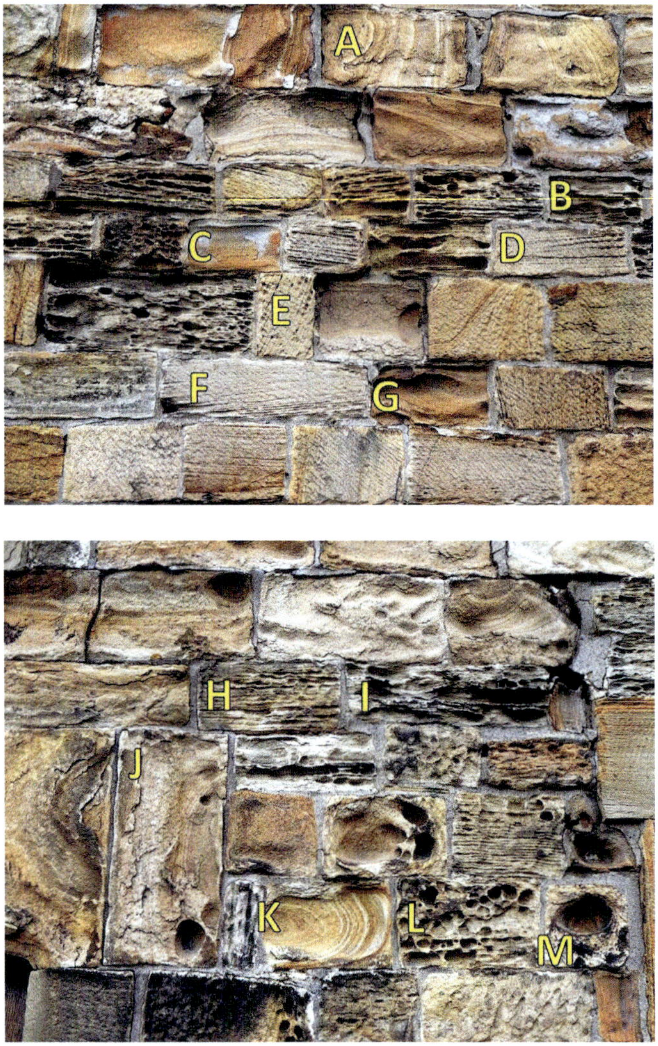

(a)

(b)

Figure 7.39 Studies of weathered sandstone in the old almshouses dating from 1666 on Palace Green adjacent to Durham Cathedral (see Fig. 7.25).

F: In this cross-bedded block, the weathering is less advanced than in Block E, apart from the lower left corner of the block, and is controlled by the primary bedding fabric.

G: The weathering in this block is controlled by the rim of an iron oxide ring.

H: There are subtle differences in the properties of the bedding laminations in this block. The resistant layers may be coarser and hence originally more permeable, so that the cementing fluids could move along them more easily, leading to more precipitation and hence to a greater resistance to weathering.

I: In this adjacent stone, the 'standout' layers are quartz veins—parallel, and sub-parallel to the bedding, see also Block B. The eroded portions occur in the sandstone which, although primarily cemented by quartz, contains clays such as kaolin and illite which make it more vulnerable to weathering.

J: The geometry of the weathered surface in this stone, i.e., the flaking of the block subparallel to its surface, appears to be due to liesegang ring iron oxide surfaces.

K: The weathering is dominated by the fabric which was caused by the deposition of concentric liesegang rings.

L: The appearance of this stone is characterised by ‘tafoni’, i.e., small, cavity-like features found in granular rocks such as granites, sandy limestones and sandstones. They are a more advanced form of pitting, termed honeycombing or caverning. This form of weathering is known as carious weathering from the Latin *cariosus*, for decaying and rotten, as in teeth.

M: The stone, immediately to the right of Block L, shows directly the influence of the more resistant iron oxide liesegang rings on the weathering—which has been restricted by the series of the rings having weathered out as a cone. Note that the smaller block immediately above Block M exhibits the same type of weathering.

From the descriptions above, the reader can study the other blocks in Figure 7.39, all of which exhibit variations on the same weathering theme, i.e., the weathering being dominated by the liesegang ring fabric, the bedding fabric or both. For example, the block directly above blocks K and L exhibits two cavities where the weathering has exploited weaknesses in the microstructure caused by the liesegang rings (and producing the macabre geometry when the picture is turned on its side). Similarly, the flaking of the block to the left of block J is caused by the same phenomenon.

So, although at first sight all these building stones seem to be weathered in different ways, in fact they have all been weathered by the same mechanism: the effects of wind, rain and temperature changes on the sandstone’s microstructure, which is controlled by the primary bedding and to a greater or lesser extent by the superimposed liesegang ring fabric.

7.3.5 The problem of replacing weathered sandstone building stones

We conclude this section on Durham Cathedral by illustrating the difficulty of upgrading a building which has suffered severe weathering of its exterior building stones. In Figure 7.40, parts of the exterior of the cathedral are shown and the phenomena associated with the weathering of this particular sandstone (as discussed in the previous Section 7.3.4) are clearly evident here. The replacement facing stones, the lighter coloured blocks in Figure 7.40, that have been used are sandstone from the Upper Carbonifereous Millstone Grit series and are fine-grained, thick-bedded, buff-brown-cream in colour and with an iron-stained speckle.

There does not seem to be any ideal method of renovating these exterior weathered cathedral walls. We commented in note 2 that in 1776 a local architect was hired to remedy the exterior decaying stonework of the Cathedral and at that time four inches (~100 mms) were scraped off the surface. No doubt this was effective in its way but there is a limit to how many times this could be done, bearing in mind the increase in vertical stress that results from reducing the cross-sectional area of the walls, not to mention the difficulty and duration of such an operation. Instead, a block-by-block replacement approach has been used which avoids the problems of the ‘scraping’ operation. However, because the replacement blocks are of a different sandstone, a different colour, and are relatively new and unweathered, they are not visually harmonious. Nevertheless, until an alternative restoration procedure is developed, it seems that such a compromise is inevitable.

(a)

(b)

Figure 7.40 Replacement of severely weathered blocks, exterior of Durham Cathedral.

NOTES

1 *Fleche* is a French term which refers to any spire. However, in English its use is more restrictive: it is used to describe a small, slender spire placed on the ridge of a church. It is usually built as a wooden framework covered with lead or occasionally copper and is generally of rich, light, delicate design, in which tracery, miniature buttresses and crockets have important parts.

2 Johnson (1980) notes that in 1776, a local architect was hired to remedy the exterior decaying stonework and that "He scraped no less than four inches [100 mms] off the surface and so removed over 1,000 tons of masonry, including virtually all the original mouldings, at a cost of £30,000."

Chapter 8

Deterioration of building stones and stone buildings

8.1 INTRODUCTION

We have already touched on the subject of building stone deterioration, both in Section 4.4 i.e., in an overall context, and in Section 7.3 on Durham Cathedral when discussing the specific mechanisms of sandstone degradation. In this chapter, we expand these discussions to note the overall factors involved in the deterioration of building stones and stone buildings, beginning in Section 8.2 with some of the specific physical and chemical mechanisms affecting limestones and sandstones. In his book *Stones of Britain* (1957), Shore notes that "Whether it be a ship or a building, a garden or a picture, the proper beauty of almost anything made by man depends upon maintenance." However, deterioration of building stones can occur even when buildings are maintained. Then, in Section 8.3, we consider the whole subject under the physics umbrella of the Second Law of Thermodynamics and entropy—this sounds complicated, but it is explained simply and it is a direct explanation of why ruins are inevitable.

An interesting and related additional aspect of the subject is the modern technique of laser scanning the surfaces of a building and hence having a digital record of its geometry—which means that in due course when the building is in a ruined state it could be 'reborn' as a new structure via digital printing. We illustrate this by the 15-m-high arch of the damaged Temple of Bel in Palmyra, Syria, which was replicated and assembled as a demonstration in Trafalgar Square, London. Then, following a note on the deterioration of religious buildings, we present a case history of analyses into the phenomenon of the 'bowing' of Carrara marble when it is used as cladding on the exterior of prestigious buildings.

8.2 THE MECHANISMS OF DETERIORATION

Weathering, physical and chemical, is the breaking down of rocks through contact with the Earth's atmosphere, water and biological organisms. *Physical weathering* of a rock is caused by mechanical processes; these include the expansion of water when it freezes within the pores of a stone causing the surface of the block to spall, with a similar effect when the stone is being subjected to thermal fluctuations. The phenomenon of water flowing through intact rock has been observed since antiquity: in Creech's (1682) English translation of Lucretius' six books of Epicurean philosophy on the nature of things, in Book I are the lines,

> Tho free from Pores, and Solid Things appear,
> Yet many Reasons prove them to be Rare:

For drops distill, and subtle moisture creeps
Thro hardest Rocks, and every Marble weeps. . . .

and the Chinese have a saying "Dishui chuanshi" meaning "Dripping water can pierce a stone."

The minerals making up a rock have different coefficients of thermal expansion and, as a result, will expand and contract by different amounts. This leads to the generation of stresses within the rock which can lead to its disaggregation. The action of wind containing stone particles, e.g., sand, can have a devastating effect on building stones; however, this process is classed as erosion rather than weathering.

Chemical weathering is caused by the chemical interaction of the atmosphere with stone. It is mainly the result of rainwater in which carbon dioxide has dissolved—forming carbonic acid—which can react with the mineral grains in the rock to form new minerals (clays) and soluble salts. Chemical weathering can be the result of natural processes, such as the dissolution of limestone and the hydrolysis of feldspars by acidic rainwater, or the result of pollution. An example of such anthropomorphically stimulated weathering has been discussed in Chapter 7, where the effect of sulphuric acid (caused by the dissolution of industrially generated sulphur dioxide into rainwater) on the building stone of Durham Cathedral and Castle was explained.

The three most common types of chemical weathering are solution, hydrolysis and oxidation. Solution of rock usually occurs as a result of the action of acidic rainwater, with limestones being particularly prone to such weathering, as is illustrated by the example of Portland stone on the exterior wall of the Bank of England (shown in Fig. 1.12), this stone having been exposed to the British weather for many years, especially the rain and the frost. Hydrolysis is the breakdown of rock by acidic water to produce clay and soluble salts; a well-known example is the breakdown of feldspars (a major constituent of igneous rocks, e.g., granite) to form clays, such as kaolin (china clay). Oxidation is related to the metals within a rock: the most commonly observed is the oxidation of iron in its ferrous (Fe^{2+}) state, which combines with oxygen and water to form hydroxides and oxides, such as goethite, limonite and hematite in which iron exists in its ferric (Fe^{3+}) state. In addition to giving the rock a reddish-brown colouration on the surface, this process may also weaken the rock so that it crumbles easily.

Physical weathering results from temperature and stress changes. We earlier (Fig. 3.18) included a photograph of rock spalling in the Gobi Desert caused by extreme day-to-night temperature changes, resulting in the rock repeatedly expanding and contracting until it eventually fractures. The combined effect of temperature changes and water causing stone fracturing occurs when there are repeated cycles of water ingress and freezing temperatures; water contained in cracks freezes, there is an increase in volume from water to ice and the stone breaks apart. Note that the two types of weathering, chemical and physical, can occur together; moreover, the processes of weathering can be complex, not only because of the physical-chemical interactive factors but also because other factors, e.g., biological, can be involved.

Also, deterioration of building stones and stone buildings is occurring all the time: the question is whether this deterioration significantly affects their utility and/or appearance and whether it is practical to repair the damage. We noted in earlier chapters and above that the deterioration of building stones is caused by natural decay, such as severe temperature fluctuations or repeated contact with moisture and wind accelerated by the presence of salts, but there is also damage caused by non-natural processes, such as cannon and undermining attack in the case of medieval castles, structural and decorative adjustments to religious buildings because of changes in preferred architectural styles and bombing during war.

An example of natural decay is the gravestone in Figure 8.1, which has unfortunately split along the stone's original bedding planes, a process known as delamination. In this case, the gravestone would have to be remade from a rock, such as a granite, which is not prone to split under the local weather conditions. An example of 'architectural deterioration' is the rebuilding of an external wall of St. Albans Abbey in England, shown in Figure 8.2, where the original medieval white Totternhoe stone arches have been assimilated into the newer structure. Although this stone is geologically part of the Chalk series, it does have reasonable resistance to the weather conditions. Another example is damage to an exterior wall of the Victoria and Albert Museum in London caused by bombing during the Second World War, Figure 8.3.

In Chapter 3 on recognising the different types of building stone, Sections 3.4 and 3.5 concern limestones and sandstones; these two types of building stone, of which there are

Figure 8.1 Delamination of a sedimentary rock, a sandstone, used for a gravestone.

Figure 8.2 Earlier white Totternhoe stone arches assimilated into the building during major structural alterations of Cathedral.

Figure 8.3 Damage caused to an exterior wall of the Victoria and Albert Museum caused by bombing during the Second World War, south end of Exhibition Road, South Kensington, London. It has been left unrepaired as a memorial to the attack.

many varieties, are by far the most widely used and so we discuss the mechanisms of deterioration with reference to these two stone types.

8.2.1 Deterioration of limestone

In a poem entitled "In Praise of Limestone" by W. H. Auden, the last lines are "what I hear is the murmur of underground streams, what I see is a limestone landscape" and indeed we are all aware that there are extensive limestone cave systems in Great Britain, for example in South Wales and Somerset—formed by running water which not only dissolves the calcium carbonate of the limestone but also deposits it in the form of stalactites and stalagmites. In earlier days, the circumstances relating to limestone building stones were exacerbated by the presence of sulphur compounds in the smoke produced by burning coal, the resultant sulphuric acid reacting with the limestone's calcium carbonate to form calcium sulphate, i.e., gypsum ($CaSO_4.2H_2O$). This chemical is not particularly soluble in water but it tends to crystallise with a 20% increase in volume, resulting in breakage of the surface layers of the building stone. This also applies to dolomitic limestones which are made of calcium-magnesium carbonate ($CaMg(CO_3)_2$). However, their reaction with the sulphuric acid produces magnesium sulphate, which is soluble in water and so the deterioration is more significant. These chemical effects can produce a white powdery growth composed of calcium, magnesium and sodium sulphates on the surface of building stones.

With the reduction of coal burning in urban areas, this type of damage is now much less than in Victorian times. However, stone buildings near the sea suffer from salt (sodium chloride) in the air and chlorine in the rain. Also, the permeability, or more strictly the hydraulic conductivity, of limestone plays an important role because the deterioration of gutters and drainpipes on old buildings can result in years of water seepage across the surfaces of the building stones. As we have noted earlier in the book, it is a sensible approach to use granite for the foundation base levels of pillars and walls together with the appropriate location of drainpipes, and maintaining them in good condition, to avoid damage induced by rainwater (see, e.g., Fig. 3.3 and the damage shown in Fig. 3.4).

8.2.2 Deterioration of sandstone

The weathering of sandstone has been discussed in some detail in Section 7.3 as a case study with reference to the building stones of Durham Cathedral, the weathering being exacerbated by an earlier polluting atmosphere. The stone's deterioration tendencies are primarily a result of the geochemical nature of the sandstone, repeated contact with moisture and wind, and accelerated by the presence of salts, e.g., hydrous calcium sulphate (gypsum). These salts expanding within the stone's pores cause a breakdown of the microstructure which further accelerates the deterioration process. As with the limestone, problems can be aggravated by the increased water being fed onto the adjacent stonework from defects in a building's gutters and drainpipes.

Lichens, symbiotic intergrowths of algae and fungi, can affect the appearance of sandstone surfaces, as in Figure 8.4, although they are most common in rural sites and do not generally have any adverse structural effect.

Figure 8.4 White lichen on a sandstone block of a railway bridge in Delamere Forest, Cheshire, England.

8.3 THE SECOND LAW OF THERMODYNAMICS, ENTROPY AND THE INEVITABILITY OF RUINS

From the dream of the dust they came
As the dawn set free.
They shall pass as the flower of the flame
Or the foam of the sea.
Marjorie Pickthall, *A Saxon Epitaph*, (1883–1922)

Majestic though in ruin.
John Milton, *Paradise Lost*, Bk. II, (1608–74)

Let us now consider the whole subject of building stone and stone building degradation within the wider scope of the laws of thermodynamics, the development of these laws being one of the most important contributions to science. The first law states that energy cannot be created or destroyed. The second law states that the entropy of a system will increase over time (entropy being a measure of disorder in a system—in our context the deterioration of building stones and stone buildings). The third law states that the entropy of a system approaches a constant value as the temperature approaches absolute zero. At first sight, we might not consider these laws to have any direct relevance to the deterioration of building stones and stone buildings, but in fact the second law provides a fundamental basis for considering the whole subject: i.e., that the disorder of a system will continuously increase.

In various chapters, we have discussed and illustrated aspects of the deterioration of individual types of building stones—how some sandstones are friable (can easily be broken up) and other stones such as the 414-million-year-old Ross of Mull granite used to construct the Skerryvore lighthouse (Section 5.2.2) are highly resistant to degradation. But all the building stones and the stone buildings will degrade in time, the associated physical and chemical processes being an inevitable result of the Second Law of Thermodynamics together with the associated concept of entropy. This means that the processes leading to the degradation of building stones and stone buildings are inevitable and irreversible (unless further energy is introduced into the system). It's just a question of time and whether any remedial measures have been introduced.

Consider the dilapidated building in the Scottish Highlands shown in Figures 8.5 and 8.6, noting that when the word 'dilapidated' is used to describe the state of the building, we are using the word literally—because it means to bring a building to ruin (derived from the Latin *lapis* for 'stone' and *dilapidatus* meaning 'to squander', that is to throw away or literally scatter like stones). Whilst the Second Law of Thermodynamics and the associated increase in entropy could be applied to the whole formation of the Earth and the geomorphological development of this Highland scenery, let us consider just the cottage/barn system for our current purposes. When it was built, stones of the local Cambrian quartzite were used, the smooth, flat faces of the stones being joint planes. This quartzite contains several sets of joints, all of which formed at right angles to the bedding and which probably formed during the uplift and exposure of the rock at the Earth's surface. The joints combine with the bedding to break the rock into approximately rectangular blocks, ideal in size and shape for building purposes. The wooden roof struts were then prepared and

Figure 8.5 Dilapidated cottage/barn in the NW Highlands of Scotland near Inchnadamph in the parish of Assynt, south-west Sutherland.

Figure 8.6 Close-up of the dilapidated stone barn in Figure 8.5.

roof slates were brought to the site. All this involved the use of energy, as did the actual construction of the barn.

At some stage, the building was abandoned and left to the 'mercy of the elements'—resulting in its current state of disrepair. The stone components have weathered well, being a hard crystalline quartzite which is extremely resistant to both weathering and erosion. However, understandably the dry stone wall sheep fold adjacent to the barn is collapsing despite the individual stone blocks remaining virtually unweathered, whereas the cemented walls of the barn are still intact. Slippage of the roof boards to which the slate tiles have been fastened has caused the structural integrity of the barn to have been lost, and it would require the input of more energy to reverse the physical effect of this degradation processes, i.e., by reducing the entropy of the barn but increasing the consequential entropy of the world.

This example, of the construction of a building, is one illustration of a local decrease in entropy at the expense of greater entropy in the surrounding environment. Arthur Eddington, the noted physicist, called the inevitable increase in overall entropy "Time's Arrow" and, in the context of this book, it applies to all the physical and chemical mechanisms that lead to the degradation of building stones and stone buildings, i.e., the inevitability of ruins. Rose Macaulay (1953) provides a positive approach to this inevitability by giving her erudite and extensive book the title *The Pleasure of Ruins*. In the introduction, she includes the sentence:

> . . . it is interesting to speculate on the various strands in this complex enjoyment [of ruins], on how much of it is admiration for the ruin as it was in its prime, how much aesthetic pleasure in its present appearance, how much is association, historical and literary, what part is played by morbid pleasure in decay . . . and by a dozen other entangled threads of pleasurable and melancholy emotion, of which the main thread is, one imagines, the romantic and conscious *swimming down the hurrying river of time* [italics added].

In addition to 'natural deterioration' of stone buildings, there can be other factors, e.g., the enacting of the 1534 Act of Supremacy in Britain through which King Henry VIII replaced the Pope in becoming responsible for 'the good order of the Church'. As a consequence, Thomas Cromwell was ordered in 1535 to arrange visitations to be made of all 'religious houses' to explain the new Injunctions, i.e., new rules. This period is known as the Dissolution of the Monasteries, the suppression of the smaller monasteries starting in 1536 and the suppression of the larger monasteries starting in 1539—the plunder of the lesser ones being so profitable that King Henry was easily induced to doom the greater to the same fate. The consequence was that the use and upkeep of religious buildings were severely disrupted and the changes led to the remodelling of country society at that time.

With reference to this period, Trevelyan (1964) concisely notes that "The crash of monastic masonry resounding through the land was not the work of the 'unimaginable touch of time' but the sudden impact of a king's command, a demolition order to resolve at one stroke a social problem that had been maturing for two centuries past." Although many religious buildings were put to another use, e.g., their framework was used for new domestic dwellings, see the book's cover and Figure 8.7(a) and (b), or as part of a new country house, other buildings such as Tintern, Whitby and Fountains Abbeys were just left to decay.

Equally significant in cities is the complete replacement of old style buildings with newer architecture having greater internal space, utility and exterior attractiveness—sometimes with a reminder of the previous buildings, as in the 'memorial' sandstone fragments in Sydney's Botanic Gardens. These are shown in the evocative Figure 8.8 against the background of active new construction. The Botanic Gardens describes this garden as "Memory is Creation Without End" and explains that it "symbolises the circular connection of past, present & future. In salvaging & reconfiguring the stones into this spiral unification of sculpture & landscape the artist endows them with new life, meaning & memory." In other words: the ever-present entropy in action.

8.4 DIGITAL RECORDING OF STONE BUILDINGS AND THE POSSIBILITY OF THEIR REBIRTH

Having noted the inevitability of ruins, modern techniques can enable buildings to be duplicated or resurrected if they have been destroyed. This is because the development of 3D cameras and laser scanning methods have enabled digital recordings to be made of the geometry of stone buildings. With this approach, building geometry data acquisition is achieved using a 3D scanner which emits light and detects its reflection from an object, thus enabling the distance to the object to be determined. The procedure has been developed since the 1990s and is now widely used in different disciplines for 3D measurement, surveying, documentation and modelling.

Figure 8.7(a) Houses built within the ruins of the arched west front of Bury St. Edmund's Abbey, see Figure 8.7(b).

Figure 8.7(b) A drawing based on several reconstructions of the original Bury St. Edmund's Abbey. Note the three large main West Front arches—into the ruins of which were built the houses shown in Figure 8.7(a) and on the cover of the book.

Figure 8.8 Replacement of old stone buildings with newer concrete buildings. This picture was taken in the Botanic Gardens in Sydney, Australia, and shows the stone fragments of the old buildings in the foreground and the new buildings being constructed in the background.

The advantages for stone building studies are wide-ranging:

- A large area/volume can be recorded at high resolution in a short time,
- The data are digital and so co-ordinates can be established for all points on the building,
- The procedure is safer than some other methods because no access is required other than locating the scanning device,
- Lighting is not required, and
- A 3D virtual image can be created with 3D co-ordinates.

(a) (b) (c) (d)

Figure 8.9 3D printed demonstration replica in London of a surviving 15-m-high arch at the Temple of Bel in Palmyra, Syria. (a) The arch in Trafalgar Square, London. (b) The arch viewed with the National Gallery in the background. (c) A closer look at part of the arch illustrating how the stone texture is replicated in the 3D printed version. (d) A close-up of part of the structure.

Needless to say, the advantages for establishing the geometrical state of deterioration of a building are many, not least of which is that a day's work of laser scanning the outside of a building enables the whole surface to be characterised in terms of the weathering features. This immediately suggests how valuable the technique is for research into building stone weathering characterisation and the urgency of remedial action for a specific building. Also, because a building can be replicated, there can be two or more of them! So, although we have used the phrase 'the inevitability of ruins' in the heading of Section 8.3, if the building or indeed the ruin of a building has been digitally recorded via laser scanning, then it can now be replicated using 3D printing techniques. This was demonstrated in 2016 when a replica of the ruined and weathered Syrian Palmyra Arch was built with 3D printed components (using stone powder and a lightweight composite) in Trafalgar Square, London, see Figure 8.9. This demonstration project was organised by the Institute for Digital Archaeology, which is based in Oxford and Harvard universities.

8.5 A NOTE ON THE DETERIORATION OF RELIGIOUS BUILDINGS

The magnitude of the problem relating to the deterioration of religious buildings is indicated by the fact that there are of the order of 16,000 religious buildings in England, of which 42 are Anglican cathedrals and most of which require maintenance and significant repairs. The problem is that the necessary finances to fund the repairs (running into hundreds of thousands of pounds) are not easily available and the members of the clergy are generally not trained in the necessary organisational and accounting skills. Some deterioration may be superficial, such as weathering of the exterior stone surfaces, but other deterioration may affect the structural integrity of the building itself. For example, in Chapter 5 we explained the purpose of the external flying buttresses in resisting the lateral force exerted by roof arches. If the stone in these flying buttresses deteriorates to such an extent that they can no longer sustain the lateral and downward forces, then they will collapse, along with part of the roof. Another factor is the unfortunate fact that the lead used for waterproofing church roofs is often stolen, considerably accelerating deterioration of the roof.

In 2017, the Taylor Review of Sustainability of English Churches and Cathedrals was published (and is freely available via the internet). This report notes that, "With 16,000 parish churches across the country, 12,200 of them nationally recognised as being of outstanding architectural and historical interest, maintenance and repair of these beautiful and important buildings represents a serious challenge. These are complex buildings, constructed using traditional skills and materials that are expensive to maintain and repair." Also highlighted is that, "Most historic buildings will, on a medium to long-term cyclical basis, require a large injection of capital funding for major works such as replacing the roof or repairing decayed stonework, even where this is guarded against by responsible and regular routine maintenance."

The irony is that, whereas in the time of Henry VIII when there was a specific decree to stop the monasteries and associated buildings from operating, taking their money and encouraging their destruction, nowadays there is a desire to mitigate the deterioration of religious buildings, but there are insufficient funds to achieve the objective.

8.6 A CASE HISTORY OF STUDIES OF THE CARRARA MARBLE BOWING PHENOMENON

Much of this book has been dedicated to the use of stone for structural purposes, i.e., as the building itself, but there is also another aspect: stone's decorative use via external cladding. We concentrate here on Carrara marble, partly because of its ubiquity and partly because of the interesting technical reasons for the unfortunate tendency of Carrara marble panels to 'bow'. All readers of this book will have seen, whether consciously or unconsciously, the white Carrara marble with its characteristic dark 'smudges'. It is used widely, internally as table tops and bathroom tiles, and externally as building cladding. Although there are different types of Carrara marble, it is the Sicilian or White Carrara (Bianco Carrara) that is the type used for external use, such as for the Marble Arch monument at the west end of Oxford Street.

One of the many Carrara marble quarries in Italy is shown in Figure 8.10 and further information on the actual quarrying procedure is given in Section 4.3.3. Unfortunately, when Carrara marble is used as external decorative cladding on buildings, it sometimes has a tendency to bow, i.e., the individual panels become curved. An example of this was the Standard Oil Building in Chicago, which used a tubular steel-framed structural system clad with 43,000 thin slabs of Carrara marble and was the fourth tallest building in the world in 1974 when it was completed. The Carrara marble cladding cracked and bowed to such an extent that the entire building had to be refaced in the early 1990s with white Mount Airy granite. Another example is illustrated by the photographs in Figure 8.11 of parts of the exterior of Finlandia Hall in Helsinki, Finland, noting that Carrara marble panels can bow either inwards or outwards.

Figure 8.10 Part of the Carrara marble quarry system in Italy. (Photo by authors and courtesy of Imperial College Press, London)

(a) (b)

Figure 8.11 Finlandia Hall in Helsinki illustrating the outward bowing of the Carrara marble panels. (Photo by authors and courtesy of Imperial College Press, London)

Finlandia Hall, designed by the Finnish architect Alvar Aalto and built between 1967 and 1972, was clad with 7000 m^2 of Carrara marble panels 30 mm thick. However, the panels soon started to bow, and in 1983 by up to 50 mm (2 inches). In addition to this problem, there was surface deterioration and cracking around the fixing points. Following debates about the way forward, in 1997 it was decided to replace the cladding with new Carrara marble panels and to make improvements, such as checking the marble strength, reducing the size of the panels and improving the panel fixing arrangement. The new cladding was in place in 1999—but the marble started bowing again; not only that but, while the original panels had bowed inwards, the newer panels bowed outwards, Figure 8.11.

Because this type of external panel bowing had been a problem for many buildings in different countries, a major European research effort was undertaken in 2000–5 to examine the likely causes. This research project was the largest to have been undertaken on building stones—as far as we are aware—and a brief overview is provided here. The project had the title of *Testing and Assessment of Marble and Limestone* and was co-ordinated by the Swedish National Testing and Research Institute (TEAM Report, 2005). Sixteen partners from nine EU countries comprised the team of stone producers, trade associations, testing laboratories and standardisation/certification bodies. In the TEAM report, it is noted that, "Although the vast majority of reported durability problems with thin marble or limestone slabs refer to the Italian Carrara marble, which is also by far the most widely used marble in the world, other marbles, e.g., American, Norwegian and Portuguese, have also been reported to bow on building façades. However, the reports on performance of Carrara marble are inconsistent, since in some cases Carrara marble apparently performs satisfactorily." They note also that there are about 200 different stone quarries in operation in the Carrara marble area and that the direction of the observed bow may be either convex (i.e., outward) or concave (i.e., inward) relative to the façade. An example of the bowing for the Danish National Bank is shown in Figure 8.12. Notice that the most severe bowing is on the southern side of the bank, indicating perhaps that the heat from the sun and the associated temperature changes could well be one of the causes.

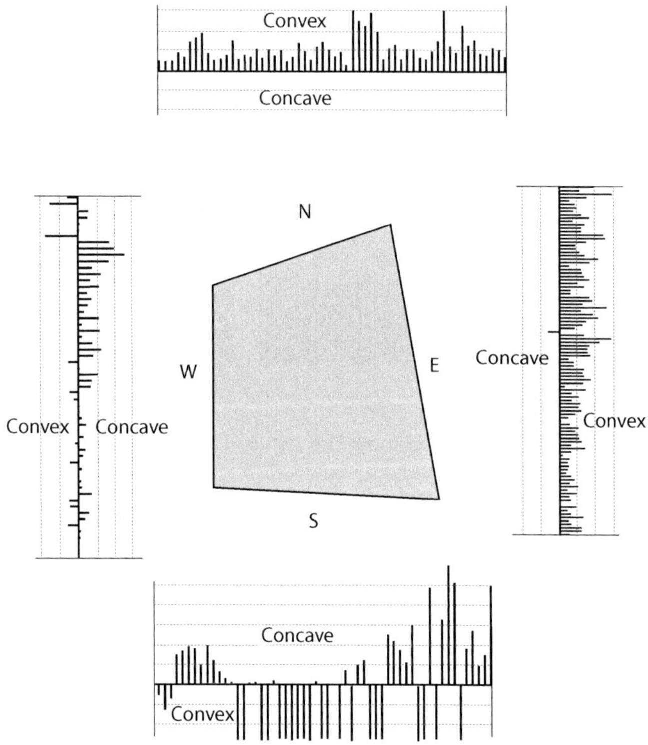

Figure 8.12 Bow measurements for individual Danish National Bank panels in 2001, second row from the ground. Scale for bowing is in steps of 1 mm/m. (Redrawn from TEAM (2005) *Testing and Assessment of Marble and Limestone, Final Technical Report*. Sweden: SP Swedish National Testing and Research Institute)

The main TEAM objectives listed in the Report were:

1 To establish a sound understanding based on natural sciences of the phenomena leading to poor field performance of marble clad façades.
2 To develop a laboratory test method for determination of potential bowing of thin slabs of natural stone.
3 To develop a field monitoring, evaluation and repair guide for façade cladding, which will include risk assessment and service life prediction.

Some suggestions for the panel bowing were 1) anisotropic (in different directions) thermal expansion of calcite and dolomite, 2) influence of moisture (and possibly free water) and temperature variations and 3) release of locked-in rock stresses. Extensive data were recorded for buildings that had suffered the bowing problem, together with the environmental conditions. However, it was not possible to provide a predictive system for the bowing from these data. Considerable research was then undertaken, including taking many samples (Fig. 8.13) and measuring the *in situ* stress in the Carrara quarries, plus the direction of cutting in the quarry, and studies of the stone's microstructure.

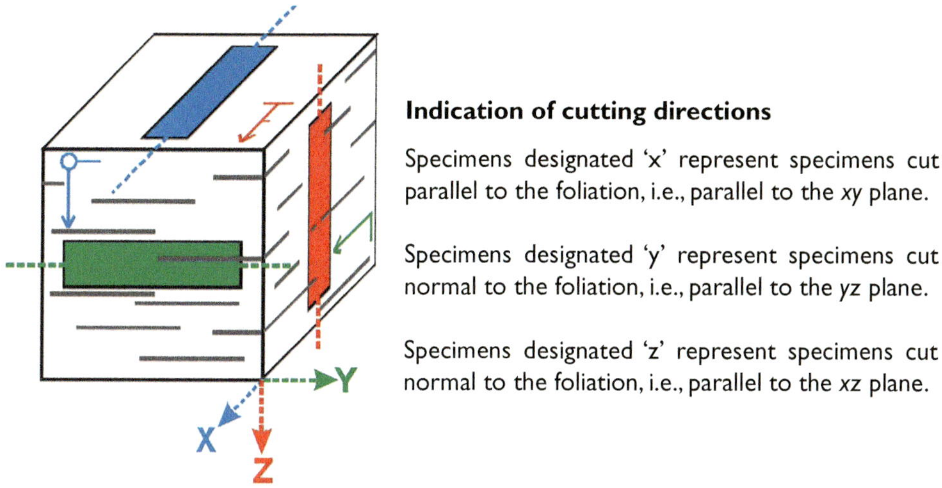

Figure 8.13 Cutting directions specification in the Carrara marble for the TEAM Project experiments. The term 'foliation' refers to the planar structure that results from flattening of the constituent grains of a metamorphic rock by natural causes. (Modified from TEAM (2005) *Testing and Assessment of Marble and Limestone, Final Technical Report*. Sweden: SP Swedish National Testing and Research Institute)

In the conclusions of the TEAM project, it was noted that the project was by far the largest research and development project ever directed towards building stones, and that the work involved a literature review of more than 300 articles, a survey of about 200 buildings and a comprehensive programme of laboratory and field work. The project had 16 partners from nine countries, a budget of about €4.2 million, and was partly financed by the European Commission. However, whilst the work involved many laboratory and field tests, plus the detailed monitoring of buildings, and the report makes many valuable conclusions regarding aspects of the Carrara marble and the environmental conditions, it was not possible for the project to make clear-cut conclusions regarding methods for predicting the marble bowing when using marble from a specific quarry to clad a specific building.

A publication by Royer-Carfagni (1999) reported on a scanning electron microscope study of samples cored from the panels of Finlandia Hall's façade. This indicated that 'granular decohesion' was the most important sign of marble damage at the microstructural level. Tschegg (2016) studied the environmental influences on damage and destruction of the structure of Carrara marble using acoustic emission testing. Citing a variety of references, he referred to three phases of the structural damage and degradation: Phase I, when the outside temperature changes significantly, leading to intergranular decohesion damaging the marble pore structure and increasing the hydraulic conductivity; Phase II, when water penetration and ice crystals further damage the pore structure; and Phase III, when this process continues over longer periods of time. Tschegg demonstrated directly with his equipment that damage to the marble is caused by temperature and water changes affecting the microstructure through cracks initiating and propagating, friction at the grain boundaries and cracks at impurities of the grain surfaces.

From the work of the TEAM project, the references cited, and the principles of structural geology and rock mechanics, the reader is invited to speculate with us on the likely reasons for the Carrara marble bowing. We know that the Carrara marble from Italy is the result of three Tertiary overprinted tectono-metamorphic deformations of a limestone. These resulted in the marble having an orthotropic structure, i.e., having different properties in the three main directions, both on the small and large scales, a feature exploited during quarrying. We also know that the *in situ* rock can contain high active stresses because rockbursts have occurred in some underground rooms at the Carrara marble quarry. We also suspect that there are small-scale residual stresses within the complex marble microstructure as a result of anisotropic expansion and contraction of single crystals, both before the marble was quarried and subsequently.

With regard to external factors, the data in Figure 8.12 illustrate the tendency for the bowing to be greater on the sunniest side, i.e., on the south side of a building. We also know that, in the case of Finlandia Hall in Helsinki, there is a large difference between the summer and winter temperatures: the mean temperature in July is 17°C and in February it is –6°C. Similarly, in the case of the earlier panel bowing problem with the Standard Oil Building in Chicago, the mean temperature in Chicago in July is 23°C and in January is –6°C, i.e., akin to the Helsinki conditions.

Thus, we can summarise that the Carrara marble used to clad Finlandia Hall in Finland is a building stone which has an irregular, crystalline microstructure, possibly containing residual stresses, which is being subjected to significant daily and yearly temperature cycles (causing continual expansion and contraction of the microstructure), plus the effects of rain and significant frosts in the winter months. The fact that all the individual marble slabs in Figure 8.11 are bowing in the same direction is interesting. Would we expect this? Moreover, when the inward bowing panels on Finlandia Hall were replaced, all the new ones bowed outwards, i.e., in the opposite direction. That phenomenon has to have been caused by a feature of the marble itself, assuming other factors such as the panel fixing brackets, were the same. Thus, the causes of the bowing of the marble panels seem to be known, at least in principle, but definitive procedures to eliminate the bowing when choosing Carrara marble for a new cladding project are still elusive.

* * * * *

For readers wishing to learn more about the processes of deterioration and the associated conservation techniques, we recommend highly the English Heritage comprehensive book *Practical Building Conservation: Stone* edited in 2012 by D. Odgers and A. Henry. This 338-page book is one of a series of books under the Practical Building Conservation umbrella heading. The chapters in the book are Materials and History of Use, Deterioration and Damage, Assessment, Treatment and Repair, Care and Maintenance, and the Special Topic: Conservation of Ruins.

Chapter 9

Concluding comments

Some books are edifices to stand as they are built,
Some are hewn stones ready to form a part of future edifices,
Some are quarries from which stones are to be split for shaping and after use.
Oliver Wendell Holmes (1809–94)

In the vocabulary of Oliver Wendell Holmes' extended metaphor in the quotation above, we hope that this book is indeed a quarry from which stones of information can be taken by the reader to build an edifice of knowledge for use when observing building stones and stone buildings 'in the field', whether in the city or countryside. Our intention has been not just to provide information in isolation but also to provide supporting principles and facts so that the information can be extended to newly observed stone structures.

The geological information in Chapter 2 was presented first because this is the basis of the subject and it applies to all natural building stones. Recognising the geological type of stone is the first step to understanding a stone's appearance and the way in which it degrades. We have concentrated on Great Britain, a region having rocks/stones from almost the whole spectrum of geological ages and rock types, thus providing many different examples. In Chapter 3, we focused on the specific types of stone, especially granites, limestones and sandstones, with many illustrative examples. This was followed in Chapter 4 by an overview of the life of a building stone from quarrying to deterioration. The latter subject is important because we generally observe the building stones some time after their emplacement and often they will have degraded.

Not only is it helpful to understand the geology of building stones themselves but also the nature of a stone building: e.g., why was it built in that particular way? Two of the factors involved are mechanics and architectural style. For this reason, in Chapter 5 we explained the mechanics, starting with pillars and walls, passing through arches and roof vaults and finishing with castles and cathedrals; and in Chapter 6 we described the evolution of architectural styles, from Saxon through to post-modern. To illustrate many of these points, in Chapter 7 we outlined the nature of two exemplary stone structures: the Albert Memorial (a cornucopia of different building stones) and Durham Cathedral (a classic of Norman architecture and, unfortunately, extreme sandstone degradation).

It is good to know that, after millennia of use, stone continues to be used in new architectural design, as demonstrated by the 169 m^3 of Jordans Whitbed stone supplied by Albion stone, and used in the façade of Number 2, St. James Market in London, see Figure 9.1. Note that many of the building stones in the façade have curved surfaces.

(a) (b)

Figure 9.1 A Jordans Whitbed Portland stone façade, part of the St. James Market development in London.

Finally, in Chapter 8, we explained that, unfortunately, all stone and all stone buildings will inevitably decay with time: the question is the speed of this decay 'back to the Earth', relative to the intended life of the building. This depends on the geological nature of the stone and so we travel full circle back to Chapter 2, concerning the geological origins of different building stones. We can, however, feel confident that there is still plenty of stone for use in buildings for millennia to come.

In Chapter 1, we noted that, while walking in a countryside environment, it is satisfying to be able to name a particular species of bird or flower that one passes, and that the same applies to the identification of the different types of building stones in the built environment. That has been the context of our primary objective and we hope that the reader's enjoyment has been thus increased. If also, the book is considered as a quarry for explanations regarding the geology, mechanics and architecture in order to develop a deeper understanding, we shall have fully achieved our objective in writing the book.

John A. Hudson and John W. Cosgrove
Department of Earth Science and Engineering, Imperial College London

References and bibliography

Abdallah, M. & Verdel, T. (2017) Behavior of a masonry wall subjected to mining subsidence, as analyzed by experimental designs and response surfaces. *International Journal of Rock Mechanics and Mining Sciences*, 100, 199–206.

Anderson, E.M. (1942) *The Dynamics of Faulting and Dyke Formation*. Oliver & Boyd, Edinburgh, UK.

Andrews, F.B. (1974) *The Mediaeval Builder and His Methods*. EP Publishing Ltd., Wakefield, Yorkshire, UK, reprinted (1976), 109p.

Arkell, W.J. & Tomkeieff, S.I. (1953) *English Rock Terms*. Oxford University Press, London, UK, 139p.

Ashurst, J. & Dimes, F.G. (1977) *Stone in Building*. The Architectural Press Ltd., London, UK, 105p.

Ashurst, J. & Dimes, F.G. (2008) *Conservation of Building and Decorative Stone*. Butterworth-Heinemann, Oxford, UK, 254p.

Attewell, P.B. and Taylor, D. (1990) Time-dependent atmospheric degradation of building stone in a polluting environment. *Environmental Geology and Water Sciences*, 16, 43–55. Available from: https://doi.org/10.1007/BF01702222

Barman, C. (1926) *The Bridge: A Chapter in the History of Building*. John Lane, The Bodley Head Ltd., London, UK, 249p.

Bates, R.L. & Jackson, J.A. (eds.) (1980) *Glossary of Geology*. American Geological Institute, Falls Church, VA, USA, 749p.

Bathurst, B. (1999) *The Lighthouse Stevensons*. HarperCollins, New York, USA, 284p.

Bayley, S. (1981) *The Albert Memorial: The Monument in Its Social and Architectural Context*. Scolar Press, London, UK, 160p.

Beasley, M. (1998) *The World Atlas of Architecture*. Chancellor Press, London, UK, 408p.

Bennison, G.M. & Wright, A.E. (1969) *The Geological History of the British Isles*. Edward Arnold & Co., London, UK, 406p.

Bezzant, N. (1980) *Out of the Rock*. William Heinemann Ltd., London, UK, 244p.

Bignell, E. (ed.) (2018) *Natural Stone Directory 2018–2019*. QMJ Group Ltd., Nottingham, UK.

Blair, J. & Cowley, J.K. (1961) *The Cathedrals of England*. W&R Chambers, London, UK, 144p.

Bond, F. (1913) *An Introduction to English Church Architecture: Volume I, from the Eleventh to the Sixteenth Century*. Oxford University Press, London, UK, 486p.

Bond, F. (1913) *An Introduction to English Church Architecture: Volume II, from the Eleventh to the Sixteenth Century*. Oxford University Press, London, UK, 986p.

Bond, F. (1913) *Gothic Architecture in England*. B.T. Batsford Ltd., London, UK, 782p.

Brabbs, D. (1999) *Abbeys and Monastries*. Weidenfield & Nicolson, London, UK, 160p.

Brabbs, D. (2001) *England's Heritage*. Cassell & Co., London, UK, in association with English Heritage; paperback edition (2003) Weidenfield & Nicolson, 416p.

Brandon, L.G., Hill, C.P. & Sellman, R.R. (1958) *A Survey of British History from the Earliest Times to 1939: Volume 1 to 1485*. Edward Arnold & Co., London, UK.

Brangwyn, F. & Sparrow, W.S. (1914) *A Book of Bridges*. John Lane the Bodley Head, London, UK, 415p.
Brown, D.J. (1993) *Bridges: Three Thousand Years of Defying Nature*. Mitchell Beazley/Macmillan, London, UK, 176p.
Butler, C. (2002) *Postmodernism: A Very Short Introduction*. Oxford University Press, Oxford, UK, 142p.
Campbell, J.W.P. & Pryce, W. (2003) *Brick: A World History*. Thames & Hudson Ltd., London, UK, 320p.
Carpenter, H.B. & Knight, J. (1929) *An Introduction to the History of Architecture*. Longman, Green & Co., London, UK, 292p.
Cassell Petter & Galpin (eds.) (1865) *Cassell's Illustrated History of England*. Volume I. Cassell Petter & Galpin, London, UK, Paris and New York, USA.
Chambers, W.A. (1862) *Treatise on the Decorative Part of Civil Architecture*. Lockwood & Co., London, UK, 336p.
Child, M. (1981) *English Church Architecture: A Visiual Guide*. B.T. Batsford Ltd., London, UK, 119p.
Clifton-Taylor, A. (1962) *The Pattern of English Building*. B.T. Batsford Ltd., London, UK, 384p.
Clifton-Taylor, A. (1967) *The Cathedrals of England*. Thames & Hudson, London, UK, 288p.
Clifton-Taylor, A. & Ireson, A.S. (1983) *English Stone Building*. Victor Gollancz Ltd., London, UK, 285p.
Cole, E. (ed.) (2002) *The Grammar of Architecture*. Bulfinch Press, Little, Brown & Co., London, UK, 352p.
Cook, G.H. (1951) *The Story of Durham Cathedral*. Phoenix House, London, UK, 51p.
Cormack, P. (1984) *English Cathedrals*. Artus Books, London, UK, 153p.
Cosgrove, J.W. & Hudson, J.A. (2016) *Structural Geology and Rock Engineering*. Imperial College Press, London, part of World Scientific Publishing Co., Singapore, 534p.
Cragoe, C.D. (2014) *How to Read Buildings: A Crash Couse in Architecture*. Bloomsbury Publishing, London, UK, 256p.
Creech. T. (1682) *Lucretius [Titus Lucretius Carus] The Epicurean Philosopher: His Six Books*. Translated by Thomas Creech. Oxford University Press, Oxford, UK, 269p.
Crossley, H. (1991) *Lettering in Stone*. The Self-Publishing Association, Upton-upon-Severn, UK, 190p.
Dafforne, J. (1878) *The Albert Memorial, Hyde Park: Its History and Description*. Virtue & Co. Ltd., London, UK, 70p. Available from: https://archive.org/stream/albertmemorialhy00daff#page/n7
Deighton, H. (1947) *The Art of Lettering*. B.T. Batsford Ltd., London, UK, 86p.
De Maré, E. (1954) *The Bridges of Britain*. B.T. Batsford Ltd., London, UK, 226p.
De Maré, E. (1964) *London's River*. The Bodley Head, London, UK, 127p.
Denison, E. & Stewart, I. (2012) *How to Read Bridges*. Bloomsbury Publishing, London, UK, 256p.
Dewe, G. & Dewe, M. (1986) *Fulham Bridge 1729–1886*. Fulham and Hammersmith Historical Society, London, UK, 147p.
Dury, G.H. (1959) *The Face of the Earth*. Penguin Books, Harmondsworth, Middlesex, UK, 237p.
Edwards, D.L. (1989) *The Cathedrals of Britain*. Pitkin Guides, Andover, UK, 160p.
Elsden, J.V. & Howe, J.A. (1923) *The Stones of London*. Colliery Guardian Co. Ltd., London, UK, 205p.
English Heritage (2012) *Practical Building Conservation: Stone*. Ashgate Publishing Ltd., London, UK, 338p.
Ericson, J.E. & Purdy, B.A. (1984) *Prehistoric Quarries and Lithic Production*. Cambridge University Press, Cambridge, UK, 140p.
Erlande-Brandenburg, A. (1995) *The Cathedral Builders of the Middle Ages*. Thames & Hudson Ltd., London, UK, 175p.
Fairbank, A. (1977) *A Book of Scripts*. Faber & Faber, London, UK, 126p.
Fearnsides, W.G. & Bulman, O.M.B. (1950) *Geology in the Service of Man*. Penguin Books, Harmondsworth, UK, 217p.

Feilden, B.M. (2003) *Conservation of Historic Buildings*. 3rd edition. Architectural Press, Routledge, London, UK, 388p.
Field, D.M. (2010) *The World's Greatest Architecture: Past and Present*. Hermes House, London, UK, 447p.
Fletcher, B. (1943) *A History of Architecture on the Comparative Method*. 11th edition. B.T. Batsford Ltd., London, UK, 1033p.
Fletcher, B. & Fletcher, B. (1905) *A History of Architecture on the Comparative Method*. 5th edition. B.T. Batsford Ltd., London, UK, 738p. Available from: https://archive.org/details/historyofarchite00fletuoft/page/n5
Foley, J.H. (1873) *The National Memorial to His Royal Highness the Prince Consort*. vii, 100p., 25 leaves of plates: ill. (some col.). John Murray, London, UK. Available from: Kensington Central Library, London, UK.
Forde-Johnston, J. (1981) *Castles of England and Wales*. Constable & Co., London, 352p.
Fry, P.S. (2005) *Castles: England & Scotland & Wales & Ireland*. David & Charles, Newton Abbot, UK, 256p.
Gardner, A.H. (1945) *Outline of English Architecture*. B.T. Batsford Ltd., London, 122p.
Gardner, S. (1922) *A Guide to English Gothic Architecture*. Cambridge University Press, Cambridge, UK, 228p.
Garfield, S. (2010) *Just My Type*. Profile Books, London, UK, 352p.
Gibberd, V. (1997) *Architecture Source Book: A Visual Guide to Buildings Around the World*. Grange Books, London, UK, 192p.
Gill, E. (1988) *An Essay on Typography*. Lund Humphries, London, UK, 133p.
Gilpin, D. (2008) *Nature's Mighty Powers: Forces in the Landscape*. The Reader's Digest Association Ltd., London, UK, 160p.
Glanville, S.R.K. (1942) *The Legacy of Egypt*. Clarendon Press, Oxford, UK, 424p.
Gooley, T. (2015) *The Walker's Guide to Outdoor Clues and Signs*. Hodder & Stoughton Ltd., Paperback, London, UK, 438p.
Gordon, J.E. (1991) *Structures or Why Things Don't Fall Down*. Penguin Books Ltd., London, UK, 395p.
Greenwell, A. & Elsden, J.V. (1913) *Practical Stone Quarrying: A Manual for Managers, Inspectors, and Owners of Quarries, and for Students*. D. Appleton & Co., New York, USA; facsimile reprint, Nabu Press (February 1, 2012), 564p.
Gregory, G. (1806) *A Dictionary of Arts and Sciences*. Volume 1. Printed for Richard Phillips, London, UK, 960p.
Gudmundsson, A. (2011) *Rock Fractures in Geological Processes*. Cambridge University Press, Cambridge, UK, 578p.
Gwilt, J. (1874) *The Architecture of Marcus Vitruvius Pollio*. Lockwood and Co., London, UK.
Gympel, J. (2005) *The Story of Architecture*. Tandem Verlag GmbH, Potsdam, 119p.
Haining, P. (1991) *The Great English Earthquake*. Robert Hale Ltd., London, UK, 219p.
Hammond, M. (2012) *Bricks and Brick Making*. Shire Publications, Oxford, UK, 32p.
Harbison, R. (2009) *Travels in the History of Architecture*. Reaktion Books, London, UK, 287p.
Harris, E. (2009) *Walking London Wall*. The History Press, Stroud, UK, 192p.
Harris, J. & Lever, J. (1966) *Illustrated Glossary of Architecture 850–1830*. Faber & Faber, Ltd., London, UK, 302p.
Harrison, J.A.C. (1982) *Old Stone Buildings*. David & Charles, London, UK, 191p.
Haslam, A. (2011) *Lettering: A Reference Manual of Techniques*. Laurence King Publishing, London, UK, 240p.
Haywood, J. (2010) *Chronicles of the Ancient World*. Quercus Publishing, London, UK, 224p.
Heyman, J. (1995) *The Stone Skeleton: Structural Engineering of Masonry Architecture*. Cambridge University Press, Cambridge, UK, 160p.
Hill, P.R. & David, J.C.E. (1995) *Practical Stone Masonry*. Routledge, Taylor & Francis Group, Abingdon, UK, 276p.

Hill, R. (2007) *God's Architect: Pugin and the Building of Romantic Britain*. Penguin Books Ltd., London, UK, 602p.

Hislop, M. (2014) *Medieval Masons*. Shire Publications Ltd., Oxford, UK, 64p.

Holberton, P. (1988) *The World of Architecture*. WH Smith, London, UK, 210p.

Hollis, L. (2012) *The Stones of London: A History in Twelve Buildings*. Phoenix, Orion Books Ltd., London, UK, 454p.

Holmes, A. (1965) *Principles of Geology*. Thomas Nelson & Sons Ltd., London, UK, 1288p.

Holmes, O.W. (1892) *Pages From an Old Volume of Life: A Collection of Essays, 1857–1881*. The Riverside Press, Cambridge, MA, USA, 352p.

Home, G. (1926) *Roman London*. Ernest Benn Ltd., London, UK, 259p.

Hopkins, O. (2012) *Reading Architecture: A Visual Lexicon*. Lawrence King Publishing Ltd., London, UK, 174p.

Howe, J.A. (1910) *The Geology of Building Stones*. Edward Arnold, London, UK; reprinted (2012) by Forgotten Books. Available from: www.forgottenbooks.org, 455p.

Hudson, J.A. & Harrison, J.P. (1997) *Engineering Rock Mechanics: An Introduction to the Principles*. Elsevier Science Ltd., Oxford, UK, 444p.

Hunnewell, J.F. (1886) *The Imperial Island: England's Chronicle in Stone*. Ticknor & Co., Boston, MA, USA, 445p. Available from: http://hdl.loc.gov/loc.gdc/scd0001.00198543081

Jenner, M. (1993) *The Architectural Heritage of Britain and Ireland: An Illustrated A–Z of Terms and Styles*. Michael Joseph, London, UK, 320p.

Johnson, P. (1980) *British Cathedrals*. Weidenfeld & Nicolson, London, UK, 285p.

Johnston, E. (1913) *Writing & Illuminating & Lettering*. 6th edition. John Hogg, London, UK, 510p.

Jones, L.E. (1969) *The Observer's Book of Old English Churches*. F. Warne & Co., London, UK, 214p.

Jope, E.M. (1964) The Saxon building-stone industry in Southern and Midland England. *Medieval Archaeology*, 8, 91–118.

Kershman, A. (2013) *London's Monuments*. Metro Publications Ltd., London, UK, 391p.

Kinahan, G.H. (1888) Granite, Elvan, Porphyry, Felstone, Whinstone, and metamorphic rocks of Ireland. Part VI in Kinahan, G.H. (1889) *Economic Geology of Ireland: A Special Issue of Journal of the Royal Geological Society of Ireland*, Vol XVIII, 514p. Available from: https://archive.org/details/journalroyalgeo00dublgoog/page/n396

Kourkoulis, S.K. (ed.) (2006) *Fracture and Failure of Natural Building Stones*. Springer, Dordrecht, The Netherlands, 592p.

Kulander, B.R., Barton, C.C. & Dean, S.L. (1979) *The Application of Fractography to Core and Outcrop Fracture Investigations*. United States Department of Energy—Morgantown Energy Technology Center. Report METC/SP-79/3.

Lawson, A. (2010) *Anatomy of a Typeface*. David R. Godine, Jaffrey, NH, USA, 428p.

Leary, E. (1983) *The Building Limestones of the British Isles*. Building Research Establishment Report, UK Department of the Environment. Her Majesty's Stationary Office, London, UK, 91p.

Leary, E. (1986) *The Building Sandstones of the British Isles*. Building Research Establishment Report, UK Department of the Environment. Her Majesty's Stationary Office, London, UK, 115p.

Liesegang, R.E. (1924) *Chemische Reaktionen in Gallerten*. Theodor Steinkopff, Dresden and Leipzig, 90p.

Macaulay, R. (1953) *Pleasure of Ruins*. Barnes & Noble Books Inc., New York, USA, 465p.

Matthews, P. (2008) *London's Bridges*. Shire Publications, Oxford, UK, 176p.

Maude, T. (1997) *Guided by a Stone Mason*. I. B. Tauris Publishers, London, UK, 176p.

Maxwell, I. (2005) *Masonry Decay: Dealing with the Erosion of Sandstone*. Historic Scotland, Technical Conservation, Research and Education Group, Edinburgh, UK, Leaflet.

Maxwell, I. (2007) *Cleaning Sandstone: Risks and Consequences*. Historic Scotland, Technical Conservation, Research and Education Group, Edinburgh, UK, Leaflet.

Melvin, J. (2005) *. . . isms: Understanding Architecture*. Herbert Press, London, UK, 160p.

Metcalfe, L. (1970) *Bridges and Bridge Building*. Blandford Press, London, UK, 95p.

Meyer, P. & Hürlimann, M. (1950) *English Cathedrals*. Thames & Hudson, London, UK, 166p.
Milman, L. (1908) *Sir Christopher Wren*. Duckworth & Co., London, UK, 361p.
Moorhouse, A.C. (1946) *Writing and the Alphabet*. Cobbett Press, London, UK, 97p.
Morgan, M.H. (2005) *Ten Books on Architecture by Vitruvius Pollio*. Digireads.com Publishing, Stilwell, KS, USA, 183p.
Morton, H.V. (1927) *In Search of England*. Methuen & Co Ltd., London, UK, Published by Penguin Books in association with Methuen (1960), 282p.
Muir, R. (1986) *The Stones of Britain*. Michael Joseph Ltd., London, UK, 288p.
Murphy, S. (1966) *Stone Mad*. Routledge & Kegan Paul, London, UK, 229p.
Natural Stone Specialist (2017) Conservation and cleaning: federation works to protect stone's vital position in the built heritage. Report in *Natural Stone Specialist*, 52(8), 28–33.
Neville, A.M. (1981) *Properties of Concrete*. Pitman Books, London, UK, 779p.
Niemeyer, N. & Wragge, P. (1922) *The Piers Plowman Social and Economic Histories*. Book IV, 1482 to 1600. George Philip and Son Ltd., London, UK, 240p.
Nixon, P. & Dunlop, D. (1998) *Exploring Durham History*. Breedon Books, Derby, UK, 189p.
North, F.J. (1930) *Limestones: Their Origins, Distribution and Uses*. Thomas Murby & Co., London, UK, 467p.
Odgers, D. & Henry, A. (eds.) (2012) *Practical Building Conservation: Stone*. Ashgate, Farnham, UK, published in association with English Heritage, 338p.
O'Neill, H. (1965) *Stone for Building*. William Heineman Ltd., London, UK, 197p.
Parker, J.H. (1850) *A Glossary of Terms Used in Grecian, Roman, Italian and Gothic Architecture*. 5th edition; 3 volumes. David Bogue, London, UK, 528p.
Parker, J.H. (1875/1986) *Classic Dictionary of Architecture: A Concise Glossary of Terms Used in Grecian, Roman, Italian and Gothic Architecture*. 4th edition revised. Cassell plc., London, UK, New Orchard Editions (facsimile) 327p.
Parker, J.H. (1888) *A Concise Glossary of Terms Used in Grecian, Roman, Italian and Gothic Architecture*. Parker & Co., London, UK, 335p.
Parker, J.H. (1898) *ABC of Gothic Architecture*. 10th edition. James Parker & Co., London, UK, 265p.
Penoyre, J. & Penoyre, J. (1978) *Houses in the Landscape*. Reader's Union by arrangement with Faber & Faber Ltd., Newton Abbot, UK, 175p.
Penoyre, J. & Ryan, M. (1975) *The Observer's Book of Architecture*. F. Warne & Co., London, UK, 190p.
Perkins, J.W., Brooks, A.T. & Pearce, A.E. McR. (1979) *Bath Stone: A Quarry History*. University College Cardiff, Cardiff, UK and Kingsmead Press, Bath, UK, 54p.
Pettus, J. (1670) *Fodinae Regales: The History, Laws and Places of the Chief Mines and Mineral Works in England, Wales and the English Pale in Ireland*. H. L. & R. B. for Thomas Basset, London, UK, 108p.
Pevsner, N. (1943) *An Outline of European Architecture*. Pelican Books, London, UK, 301p.
Pevsner, N. (2016) *Pevsner's Architectural Glossary*. Yale University Press, London, UK, 144p.
Pevsner, N. & Cherry, B. (ed.) (1985) *The Buildings of England: London, Part 1*. Penguin Books Ltd., London, UK, 756p.
Powys, A.R. (1929) *Repair of Ancient Buildings*. J. M. Dent & Sons Ltd., London, UK, 208p.
Přikryl, R. & Smith, B.J. (eds.) (2007) *Building Stone Decay: From Diagnosis to Conservation*. Geological Society London, London, UK, Special Publications, 271, 330p.
Prior, E.S. (1905) *The Cathedral Builders in England*. Seeley & Co., Ltd., London, 112p.
Pudney, J. (1972) *Crossing London's River*. J. M. Dent & Sons, London, UK, 176p.
Rainsford-Hannay, F. (1957) *Dry Stone Walling*. Faber & Faber Ltd., London, UK and 3rd reprint (1999) South West Scotland Branch of the Dry Stone Walling Association, 129p.
Reid, R. (1980) *The Book of Buildings: A Traveller's Guide*. Michael Joseph Ltd., London, UK, 447p.
Rice, M. (2009) *Rice's Architectural Primer*. Bloomsbury Publishing, London, UK, 240p.
Rice, M. (2013) *Rice's Church Primer*. Bloomsbury Publishing, London, UK, 224p.

Roberts, C. (2005) *Cross River Traffic*. Granta Books, London, UK, 196p.

Roberts, M. (2013) *The Buildings and Landscapes of Durham University*. Profile Books Limited, Durham, UK, 160p.

Robinson, A. (2007) *The Story of Writing*. Thames & Hudson, London, UK, 232p.

Robinson, E. (1987) A geology of the Albert Memorial and vicinity. *Proceedings of the Geologists' Association*, 98(1), 19–37. Available from: https://doi.org/10.1016/S0016-7878(87)80015-9

Rodriguez-Navarro, C., Doehne, E. & Sebastian, E. (1999) Origins of honeycomb weathering: the role of salts and wind. *Geological Society of America Bulletin*, August 1999, 1250–1255.

Rogers, P. (2008) *The Beauty of Stone: The Westminster Cathedral Marbles*. Oremus, Westminster Cathedral, London, UK, 114p.

Royer-Carfagni, G. (1999) Some considerations on the warping of marble facades: the example of Alvar Aalto's Finlandia Hall in Helsinki. *Construction and Building Materials*, 13, 449–457.

Rudgley, R. (2000) *Secrets of the Stone Age*. Random House, London, UK, 195p.

Rutter, F. (1923) *The Poetry of Architecture*. Hodder & Stoughton Ltd., London, UK, 189p.

Salzman, L.F. (1952) *Building in England*. Oxford University Press, Oxford, UK, 629p.

Schaffer, R.J. (2004) *The Weathering of Natural Building Stones*. Donhead Publishing Ltd., Shaftsbury, UK, 149p.

Schofield, R. (2009) *Vitruvius: On Architecture*. Penguin Classics, Penguin Group, London, UK, 440p.

Scigliano, E. (2005) *Michelangelo's Mountain: The Quest for Perfection in the Marble Quarries of Carrara*. Free Press, Simon & Schuster, Inc., New York, USA, 368p.

Scott, G.G. (1874) *Handbook to the Prince Consort National Memorial*. Published by authority of the executive committee. John Murray, London, UK, 79p. Available from: https://archive.org/details/handbooktoprince00unse/page/n5

Scott, G.G. (1879) *Personal and Professional Recollections*. S. Low, Marston, Searle, & Rivington, London, UK, 504p. Available from: https://archive.org/details/personalprofessi00scotiala

Seaby, A.W. (1925) *The Roman Alphabet and Its Derivatives*. B.T. Batsford Ltd., London, UK, 75p.

Seward, A.C. (1945) *Geology for Everyman*. Cambridge University Press, Cambridge, UK, 312p.

Sheppard, E.J. (1982) *Ancient Athens*. Then & There Series. Longman Group, Harlow, UK, 105p.

Sheppard, F.H.W. (ed.) (1975) *Survey of London: Volume 38, South Kensington Museums Area*. British History Online, London, UK. Available from: www.british-history.ac.uk/survey-london/vol38

Shore, B.C.G. (1957) *Stones of Britain*. Leonard Hill [Books] Ltd., London, UK, 302p.

Siddall, R. (2013) Building stones in the City of London: a walk from Bank underground station to the Guildhall and Gresham street. *Urban Geology in London*, No. 6, May 2015, v.4.1. Available from: www.ucl.ac.uk/~ucfbrxs/Homepage/walks/Guildhalll&GreshamSt.pdf

Simmons, J. & Thorne, R. (2013) *St. Pancras Station*. Historical Publications Ltd., London, UK, 184p.

Simple, T.L. (2010) *Alveolar Erosion and Its Conservation: Recommendations for the Sandstone Masonry at Durham Castle*. Masters Thesis, University of Pennsylvania, Philadelphia, PA, USA, 122p.

Smith, F. (1909) *The Stone Ages in North Britain and Ireland*. Blackie & Son Ltd., London, UK, 377p.

Solé, R. & Valbelle, D. (2002) *The Rosetta Stone*. Profile Books, London, UK, 184p.

Souden, D. (1997) *Stonehenge: Mysteries of the Stones and Landscape*. Collins & Brown Ltd., London, UK in association with English Heritage, 160p.

Stalley, R. (1999) *Early Medieval Architecture*. Oxford History of Art series. Oxford University Press, Oxford, UK, 274p.

Stanier, P. (1995) *Quarries of England and Wales: An Historic Photographic Record*. Twelveheads Press, Truro, UK, 120p.

Stanier, P. (2009) *Quarries and Quarrying*. Shire Album 134. Shire Publications Ltd., Oxford, UK, 32p.

Statham, H.H. (1912) *A Short Critical History of Architecture*. B.T. Batsford Ltd., London, UK, 586p.

Stevenson, A. (1848) *Account of the Skerryvore Lighthouse*. Adam & Charles Black, North Bridge, Edinburgh, UK; Longman & Co., London, UK, 472p. Available from: https://archive.org/details/accountofskerryv1848stev/page/n9

Stone, S. (ed.) (2000) *Font: Sumner Stone, Calligraphy, and Type Design in a Digital Age*. Edward Johnston Foundation and Ditchling Museum, Ditchling, Sussex, UK, 63 p.

Stow, J. (1618) *The Survay of London*. Printed by George Purflower, East End of Christchuch, London, UK, 983p.

Stratton, A. (1937) *The Styles of English Architecture. Part I: The Middle Ages*. B.T. Batsford Ltd., London, UK, 35p.

Stratton, A. (1949) *The Styles of English Architecture. Part II: Tudor & Renaissance*. B.T. Batsford Ltd., London, UK, 42p.

Swaan, W. (1984) *The Gothic Cathedral*. Omega Books, Ware, UK, 328p.

TEAM (2005) *Testing and Assessment of Marble and Limestone, Final Technical Report*. SP Swedish National Testing and Research Institute, Sweden, 133p. Available from: http://buildingteam.extweb.sp.se/PDF/TEAM%20Final%20Report.pdf

Tibbs, R. (1970) *King's College Chapel, Cambridge*. Terence Dalton Ltd., Lavenham, Suffolk, UK, 96p.

Trevelyan, G.M. (1964) *Illustrated English Social History*. Volume 1 of 4 Volumes. Penguin, Harmondsworth, Middlesex, UK, 328p.

Trigueros, E., Cánovas, M., Muñoz, J.M. & Cospedal, J. (2017) A methodology based on geomechanical and geophysical techniques to avoid ornamental stone damage caused by blast-induced ground vibrations. *International Journal of Rock Mechanics & Mining Sciences*, 93, 196–200.

Tschegg, E.K. (2016) Environmental influences on damage and destruction of the structure of marble. *International Journal of Rock Mechanics and Mining Sciences*, 89, 250–258.

Turkington, A. & Paradise, T.R. (2005) Sandstone weathering: a century of research and innovation. *Geomorphology*, 67, 229–253.

Tzonis, A. & Giannisi, P. (2004) *Classical Greek Architecture*. Flammarion, Paris, 277p.

Ulusay, R. (ed.) (2015) *The ISRM Suggested Methods for Rock Characterization, Testing and Monitoring: 2007–2014*. Springer, Cham, UK, 293p. Available from: https://doi.org/10.1007/978-3-319-07713-0

Van Lemmen, H. (2006) *Coade Stone*. Shire Publications, Oxford, UK, 48p.

Van Soest, R.W.M., Boury-Esnault, N., Vacelet, J., Dohrmann, M., Erpenbeck, D., De Voogd, N.J. et al. (2012) Global diversity of sponges (porifera). *PLoS ONE* [Online] 7(4), e35105. Available from: https://doi.org/10.1371/journal.pone.0035105

Vivian, J. (1975) *Building Stone Walls*. Garden Way Publishing, Charlotte, VT, USA, 76p.

Wacher, J. (1974) *The Towns of Roman Britain*. Book Club Associates by arrangement with B.T. Batsford Ltd., London, UK, 460p.

Ward, J. (1911) *The Roman Era in Britain*. Methuen, London, UK, 289p.

Waters, T. (1989) *Bridge by Bridge Through London*. Pryor Publications, Whitstable, Kent, UK, 80p.

Watkin, D. (1990) *English Architecture*. Thames & Hudson Ltd., London, UK, 216p.

Watkin, D. (2000) *A History of Western Architecture*. 3rd edition. Laurence King Publishing, London, UK, 704p.

Watson, J. (1911) *British and Foreign Building Stones: A Descriptive Catalogue of the Specimens in the Sedgwick Museum, Cambridge*. Cambridge University Press, Cambridge, UK, 483p.

Wheeler, R.E.M. & Wheeler, T.V. (1936) *Verulamium: A Belgic and Two Roman Cities*. Reports of the Society of Antiquaries of London, Oxford, UK, Number XI, 297p.

Whittow, J. (1992) *Geology and Scenery in Britain*. Chapman & Hall, London, UK, 478p.

Williams, A. (1913) *The Romance of Modern Engineering*. Seeley, Service & Co. Ltd., London, UK, 377p.

Williamson, T. (2012) *Inigo's Stones: Inigo Jones, Royal Marbles and Imperial Power*. Matador/Troubador Publishing, Leicester, UK, 320p.

Wilson, R.J.A. (1975) *A Guide to the Roman Remains in Britain*. Constable & Co., London, UK, 365p.

Wynne, G. (1930) *Architecture*. Thomas Nelson & Sons, London, UK, 131p.

Yarwood, D. (1976) *The Architecture of Britain*. B.T. Batsford Ltd., London, UK, 276p.

Yüzer, E., Ergin, H. & Tuğrul, A. (eds.) (2003) *IMBS 2003: Industrial Minerals and Building Stones: Proceedings of the International Symposium of the International Association for Engineering Geology and the Environment, 15–18 September 2003, Istanbul, Turkey*. Kelebek & Grafika Grup, Istanbul, 930p.

Index